SpringerBriefs in Applied Sciences and Technology

PoliMI SpringerBriefs

T0183511

Series Editors

Barbara Pernici
Stefano Della Torre
Bianca M. Colosimo
Tiziano Faravelli
Roberto Paolucci
Silvia Piardi

For further volumes:
http://www.springer.com/series/11159
http://www.polimi.it

Giuseppe Andreoni · Massimo Barbieri
Barbara Colombo

Developing Biomedical Devices

Design, Innovation and Protection

**POLITECNICO
DI MILANO**

Giuseppe Andreoni
Department of Industrial Design, Art
and Architecture
Politecnico di Milano
Milan
Italy

Massimo Barbieri
Barbara Colombo
Technology Transfer Office
Politecnico di Milano
Milan
Italy

ISSN 2282-2577   ISSN 2282-2585   (electronic)
ISBN 978-3-319-01206-3   ISBN 978-3-319-01207-0   (eBook)
DOI 10.1007/978-3-319-01207-0
Springer Cham Heidelberg New York Dordrecht London

Library of Congress Control Number: 2013944990

Printed on acid-free paper

Springer is part of Springer Science+Business Media (www.springer.com)

Preface

Innovation capability is one of the main indicators of the economic "health" and power of a nation and of a society. But inventive ability without its protection and exploitation is sterile. The success of the whole process can be achieved following a methodologically structured pathway, from the idea to the product. The knowledge about tools, strategies, time and cost is fundamental to identify the best solution. Too often this knowledge is absent in the inventors, whoever they are, both university researchers or industrial developers.

This book wants to address this lack of information providing a short but hopefully comprehensive description of the Intellectual Property forms of protection, and the methods and tools to be used in its management.

The protection of the innovation is the first choice that can also drive the exploitation. A correct first step is crucial and starts with the identification of the most suitable typology of Intellectual Property Rights (IPRs) to apply for. Coaching the innovators in this decision is the goal of this book.

It is addressed to different readers: Students, University Researchers and Professors, Designers and Industrial Researchers too, from Engineers in any discipline (mechanics, electronics, mathematics, …) to Designers and Architects. Also, all the results of innovation should be considered: products, systems, fashion items, Information and Communication Technology (ICT), etc.

Each one of these applications can need different solutions to face their protection. But before protection it is necessary to properly understand what to protect and if it is possible. This means to codify the knowledge in the suitable form (mathematical formula, electronic circuit, drawings, …) and to carry out a dedicated research into the corresponding databases and through the most proper codes and keywords. From this crosscheck arises the possibility to protect the creation in terms of patent, utility model or industrial design. Together with this opportunity and decision, the analysis of the options (for example and simply the designation of the territory where to apply for the protection, i.e. the national or international geographical area) and the related time scheduling and corresponding expenses is another key factor to consider. This leads to a reasonable and sustainable business plan that will drive the next steps of exploitation (licensing, spin-off/start-up company initiatives, …).

At the beginning we defined Innovation as an index of the economic health of a society. Health is a human basic value, and for this reason we chose Biomedical devices used in healthcare as the case study in this book.

The text was updated in spring 2013: for new or more recent laws or determinations please consider also further readings.

Giuseppe Andreoni

Contents

Chapter 1
Introduction

The protection and the exploitation of intellectual property are two of the main strategies to promote the renewal of the economic and entrepreneurial society. The expression "protection of Intellectual Property Rights" refers to the frame of rights, relating to:

- personal or moral right to be recognized as the author or creator of the work or the technical solution or brand, which is a personal and inalienable right;
- economic right related to the exploitation of the result of his/her creative activity, which is an available and transferable right.

By their very different nature and according to the rules defining them, the outcomes of human genius can be classified into three main categories:

- the results deriving from intellectual creativity, belonging to the world of art and culture (literature, organizational charts, theater and television shows, photographs, paintings, architectural projects, etc.), which are protected by those rules that are generally called copyright;
- those representing the distinctive marks or signs, such as brand, company, designations of origin, whose form of protection is registration;
- technical innovation and design, which relate to inventions, industrial designs, plant varieties, which are referred to the laws governing patent right.

Only about the last two categories, i.e. the intellectual property belonging to the world of science and technology, we can speak more properly of IPRs.

Today innovation is becoming more and more the crucial element for development and growth in all sectors of the economy. Thus the creative ability and the actions for the protection of inventive activity represent key factors for maintaining and increasing the level of competitiveness of an economic system. For this reason, in the last years we assisted to an "explosion" of patent application, in particular in those countries whose economy is rapidly growing (China, India, Brazil, ...). This phenomenon paired with a consolidation of IPR in the "traditional" western countries.

Universities and research institutes, but also multinational companies with large Research and Development (R&D) laboratories/departments, are the main sites of innovation. Feeding interaction and knowledge transfer between universities,

G. Andreoni et al., *Developing Biomedical Devices*, PoliMI SpringerBriefs,
DOI: 10.1007/978-3-319-01207-0_1, © The Author(s) 2014

research institutes, companies and institutions on one hand, and promoting the dissemination of innovative tools and services into the productive society on the other one, can fulfill the most urgent need of technology and demand of innovation which companies have to answer market requests.

Industrial property is a strategic factor that may provide opportunities for universities and companies to protect their knowledge but above all to identify key new technologies and/or technological and industrial partners with which to develop new business initiatives. However, the patenting process is often a difficult path: sometimes this discourages companies from taking care of that, thus there is a lot of innovation that is lost or is not 'properly coded' and therefore not economically exploited.

Creativity and innovation in the EU are part of a uniform system of protection of intellectual property rights ranging from industrial property rights to copyright of authors. The protection of these rights also implies that they must be protected against piracy, illegal trade and counterfeiting. The fundamental principles of the internal market (free movement of goods and services and free competition) is based in particular on the uniformity of IPR on a European scale.[1]

The protection of intellectual property is subject to a number of international conventions whose implementation is mainly carried out by the World Intellectual Property Organization (WIPO) and the World Trade Organization (WTO). Similarly and for these purposes, the European Union has founded two important institutions, namely the Office for Harmonization in the Internal Market (OHIM), responsible for the registration of Community trademarks and Community designs, and the European Patent Office (EPO). At present, the Commission is engaged in the effective implementation of a Community patent system, harmonized and more effective, legally capable of ensuring the competitiveness of European industry. On 13 April 2011, the European Commission proposed a "regulation implementing enhanced cooperation in the area of the creation of unitary patent protection", that has now to be voted by the European Parliament and the Council of the European Union. This regulation is the last step in a series of attempts—that have failed for about forty years—to set up a common patent with validity in all Member States of the European Union (EU). The European Parliament voted for the unitary patent regulation on Tuesday, 11 December 2012.[2]

In this text we want to provide the information necessary to understand what elements and what tools are available and provided for by the law in order to protect the "industrial property", i.e. the intangible value of knowledge also related to patents and trademarks. In particular we will refer to the biomedical technology as strategic field of application. Indeed, about 90 % of innovation is produced in two sectors: military industry and medicine (and related disciplines). In the last decades, luckily, medicine asserting itself as the main one.

---

[1] http://europa.eu/legislation_summaries/internal_market/businesses/intellectual_property/index_it.htm

[2] http://unitary-patent.eu/http://unitary-patent.eu/

Chapter 2
Emerging Issues in Healthcare

2.1 Medicine and Technology in Healthcare Services

We live in a world pervaded by technology. Smart systems have invaded our spaces to provide us with new services and more comfort. Healthcare has also become technologically dependent in several applications: from diagnosis to therapy and telemedicine services, all of these processes require technology—and specifically biomedical technology—to be carried out. Indeed, one of the first field of exploration of how to apply technological advances is human health.

Biomedical technologies are in general medical equipment used to diagnose and treat various diseases, ranging from simple devices to complex systems. As internationally agreed, the biomedical instrumentation can be classified according to its use, i.e. the purpose it is used for. Thus we have:

- diagnostic devices;
- therapeutic devices;
- rehabilitation devices.

Diagnostic biomedical instrumentation includes all the equipments used for diagnosis. It consists of a very wide category of devices, ranging for laboratory machines (e.g. for the analysis of blood samples), to complex systems for multidimensional and multimodal bio-image recording, and also very simple items such as a thermometer. In the last years, this category was "populated" by a lot of systems for collecting biomedical images and specifically most of the equipment used in nuclear medicine and radiology using imaging techniques for diagnostic purposes. Among the most representative examples we can mention X-ray radiography, computed tomography (CT), magnetic resonance imaging (MRI), positron emission tomography (PET), computed tomography; single photon emission tomography (SPECT), and also echography in all its several applications. Also bioelectrical potential recording devices (Electrocardiography—ECG or EKG, electroencephalography—EEG, electromyography—EMG, electroneurography—ENG, electrooculography—EOG) are another very broad and widespread class of instruments belonging to this category (Fig. 2.1).

G. Andreoni et al., *Developing Biomedical Devices*, PoliMI SpringerBriefs,
DOI: 10.1007/978-3-319-01207-0_2, © The Author(s) 2014

Therapeutic Biomedical instrumentation are all those devices, either electrical or mechanical, which support the therapeutic activity of the patient. Examples of therapeutic systems include pacemakers, artificial heart valves, the defibrillator, the neuro-stimulators, hearing aids, etc.. Usually these devices are to be kept under constant control, since they come into direct contact with the patient, either by directly interacting with him/her or by changing some physiological and/or physical parameters (Fig. 2.2).

Rehabilitation instrumentation are all the devices supporting or implementing a functional recovery. These devices are usually all the systems for physical therapy, ranging from simple proprioceptive and balance boards to the most recent robotized systems for neuromuscular training. They can be temporary solutions but also permanent prosthesis (i.e. systems that try to fully integrate in metabolic processes and/or although mechanical, can sometimes remain permanently in the body

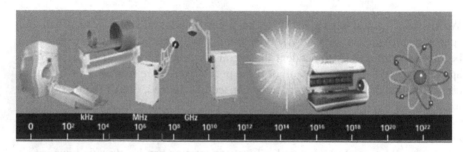

Fig. 2.1 Examples of biomedical imaging systems classified in function of the radiation used and its wavelength

Fig. 2.2 Example of therapeutic biomedical instrumentation: a pacemaker and its catheter-electrode

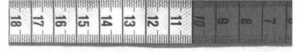

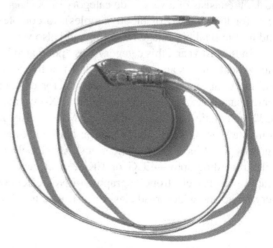

of the host, or other times can be reabsorbed by the body) or assistive technologies (wheelchairs, communication aids, and similar items so on) are part of this category (Figs. 2.3, 2.4, 2.5, 2.6).

2.2 Innovation in Biomedical Technology

Over the last 20 years we witnessed an extraordinary evolution and development of biomedical systems. Thanks to the technological availability of new solutions for sensing, computing and processing physiological data, the diagnostic capability has reached a very fine resolution both in images and in chemical/biological as well as biomechanical parameters. An example of these last systems are the motion capture techniques applied to human movement analysis. Gait analysis (Fig. 2.7) is the best known application.

It is an analysis of each component of the phases of ambulation and it is an essential part of the diagnosis of various neurologic disorders and the assessment

Fig. 2.3 Example of a balance board for physical rehabilitation of the ankle joint and for equilibrium disorders training exercises

Fig. 2.4 Example of assistive technologies: a wheelchair and its anti-decubitus cushion

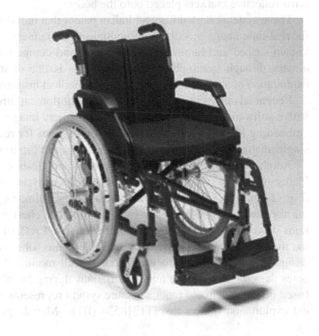

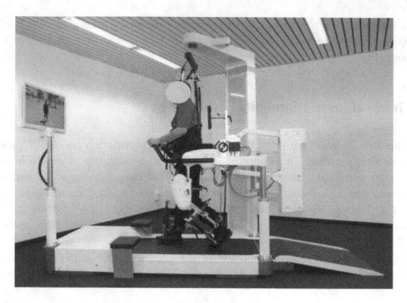

Fig. 2.5 Example of rehabilitative robot for gait training

of patient progress during rehabilitation and recovery from the effects of a neurologic disease, a musculoskeletal injury or a disease process, or amputation of a lower limb. Optoelectronic systems allowed for an automated fine resolution of human locomotion through a multi-camera setup recording the 3D positions of retro-reflective markers placed onto the body.

These systems started from an Italian patent that introduced an innovative method for real-time image processing of multiple video-camera images to detect objects of known shaped and luminosity in the scene and compute their 3-dimensional coordinates through stereo-photogrammetry. This is one of the successful example of exploitation of a patented technology in biomedical instrumentation [9] (Fig. 2.8).

Recent advances introduced markerless motion capture systems based on innovative software algorithm processing multicamera images, or wearable inertial unit embedding MEMS accelerometers and gyroscopes for recording the kinematics of single anatomical districts or more complete setup (up to total body) (Fig. 2.9).

Another recent but rapidly spreading application is the Opto-Electronic Plethysmography (OEP) for the functional analysis of the respiratory system (Fig. 2.10). The OEP system allows studying pulmonary ventilation and assessing the mechanics of breathing by measuring the chest wall volume and its variations during respiration. A large number of small reflective markers are placed on the thoracic-abdominal surface by hypoallergenic adhesive tape. A set of specially designed video-cameras analyze the chest wall motion. A dedicated software computes the enclosed volume and its variations during breathing. Also this application based on optoelectronic motion capture system represents a successful story of a patent exploitation (Patent no. IT1318534 (B1)—Metodo per misurare caratteristiche

Fig. 2.6 Example of a robotized orthosis for gait restoration: a bionic walking assistance system, composed by powered leg attachments to enable paraplegics to stand upright, walk, and climb stairs

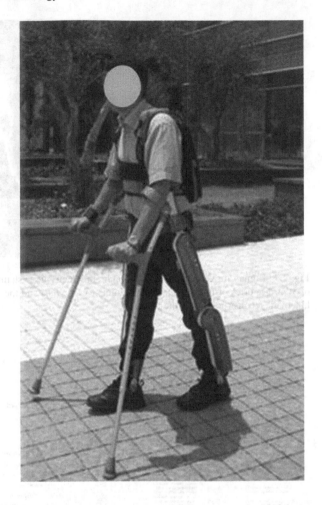

meccaniche totali e regionali del sistema respiratorio umano. Inventor(s): Aliverti Andrea; Dellacà Raffaele; Pedotti Antonio, Applicant: Milano Politecnico, Italy [9]).

All these technologies and applications are called functional analysis representing a description of a general motor and/or physical and/or physiological function. These are successful examples of different IPR typologies (technological invention, method/process invention) that have been industrially exploited.

2.3 A New Scenario

Innovation in biomedical technology now offers a wide range of solutions for the early diagnosis and effective treatment of many diseases. In modern medicine, the ICT (Information and Communication Technology) increases the possibilities of diagnosis

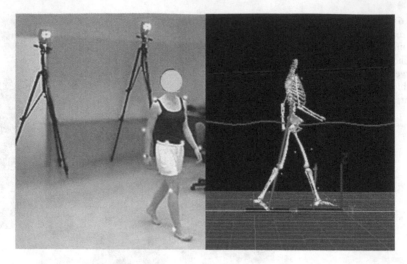

Fig. 2.7 Human movement analysis through optoelectronic system: example of data acquisition of 3D trajectories of marked anatomical points (on the *left*) and the corresponding 3-dimensional kinematic reconstruction (on the *right*)

Fig. 2.8 Human movement analysis through optoelectronic system: the first patent about the method for real-time image processing (on the *left*) and further improvements with designed application (on the *right*)

through timely communication and access to information about the state of health of the individual citizen to support the personal care. The use of mobile communications to support the development of new health services is a globally growing trend.

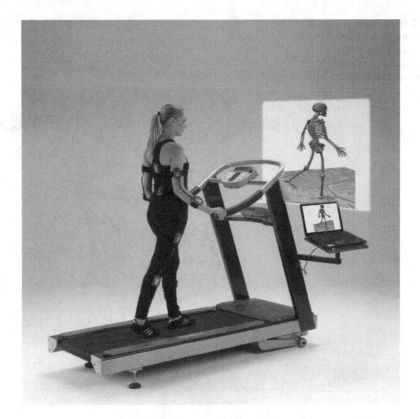

Fig. 2.9 Human movement analysis through a system of wearable inertial sensors network: example of data acquisition of the kinematics of anatomical districts and the corresponding 3-dimensional kinematic reconstruction (in the display of the figure)

Mobile Healthcare or mobile or m-Health is a term recently coined to identify medical services supported by movable type terminals such as mobile phones, smartphones, Personal Digital Assistants (PDAs), tablets, or devices that rely on wireless communication. It also includes the use of biomedical devices for wireless remote monitoring of certain physiological parameters of patients [1]. So m-Health includes reducing the use of telecommunications and multimedia technologies integrated within increasingly mobile and wireless systems to provide clinical services and health care. It may be defined as the integration of "mobile computing", biomedical sensors, and communication technology to provide new services for the health and well-being. It promises economic efficiency and improvement of the quality in the health system, with an incredible number of possible services definitely looking further than simply beyond the simple function of appointment call. Extension of mobile broadband and the promise of high-speed data transfer make the application of monitoring and remote diagnostics in near future of value both for the institutions and for the citizen.

Indeed, the convergence of technologies into mobile devices for the development of services in the most various sectors represents the platform also for many

Fig. 2.10 3-dimensional
kinematic reconstruction of
the rib cage compartments
for the implementation
of the optoelectronic
plethysmography technique

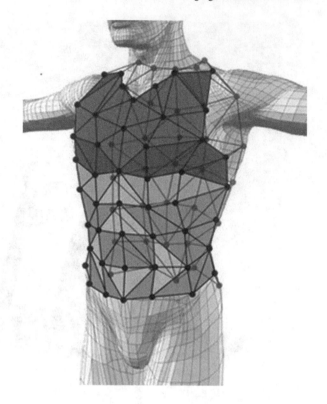

health aspects. Moreover, being health a primary need, the scenario of the devices
and the so-called "Apps" related to the health and well-being of the person represents a market of exponential growth in the short term. Mobile Health has actually
been identified in the USA as one of the main trends in health for the year 2010
[4]. Most citizens of Western countries own and use a cell phone, and today many
people use mobile systems to access services or medical information online [4].

According to Allied Health World there are currently more than 40,000 medical apps available for smartphone and tablet use and there have been 247 million
downloads of health apps in 2012, which is twice the number of the previous year
(124 million in 2011). According to the same source, 62 % of mobile users have
used their phones to research health information. Of those, 29 % are in the 18-29
age group and 18 % in the 30–49 age group, which (no surprise) shows that the
younger generations are more comfortable with and reliant on the online experience. Furthermore, it is estimated that 20 % of the online health population has
posted a health-related comment or content (which is in line with the Pareto principle: 80 % listening—20 % talking) but, according to the same source, mobile
web access increases participation in the conversation. It becomes apparent that
the mobile (as part of the online) experience is becoming ubiquitous.

Furthermore, healthcare providers benefit as much as patients from using
mobile apps: physicians are able to service patients without physically meeting

with them, reducing administrative costs by 24 %, informed decision making anywhere anytime, improve education amongst patients and students. Indeed, 94 % of healthcare providers believe mobile apps have the potential to improve care delivery and 60 % use mobile healthcare apps for work. Finally, according to some healthcare professionals, the smartphone and tablet are tending to be the most popular devices since the stethoscope.

m-Health and Telemedicine concepts rely on the medical information exchange anytime anywhere. In the last five years the same concept has been exploited by the so-called "Cloud Computing" [2]. Cloud Computing is a technology that uses the internet and central remote servers to maintain data and applications. Cloud computing allows consumers and businesses to use applications without installing them and to access their personal files from any computer with internet access. This technology allows for much more efficient computing by centralizing data storage, processing and band-width. An immediate field of application of this methodology and technology is obviously healthcare [3, 4]. Making Cloud services available to health and life science organizations is a dramatic step beyond email and communications, meetings and collaboration. They encompass a broad range of possibilities that can help you meet your organizational and ICT goals: online health and wellness tools, application development, data and image storage and sharing. Data security and patient privacy, of course, are fundamental requirements to provide reliable, secure, and cost-effective infrastructure and platform services, enabling the deployment of health science solutions and applications that typically require high availability, scalable expansion, and high performance. All Cloud Computing providers are tackling these issues and are now offering such kind of solution [5–7]. As they seek ways to optimize their technology foundation to help control costs and improve access to information, tools, and resources, health and life science organizations, both public and private, are exploring the promise of cloud computing services. Combining security and reliability with a flexible range of deployment options, cloud computing offers benefits that can make it a compelling choice for your organization. This allows new services to be built by health organizations to serve their patients and partners, and at the same time cloud computing provides a way to reduce costs, simplify management, and improve services in a safe and secure manner.

As a society, we're living longer and better than at any time in history. In part, this is due to pharmacological advances. Death rates are down, long-term disability is down, life expectancy is higher than ever, and we're making progress against the most serious diseases we face. As we consider the direction health care will take in the future, one thing is certain: new, innovative medicines will assume an increasingly prominent role in the way we improve the quality of care for future generations.

One critical role of new medicines will be the prevention, treatment, and management of many diseases suffered by an aging Baby Boomer generation. In the year 2000, there were roughly 35.6 million Americans aged 65 and older. By 2030, this number is projected to double to an estimated 71.5 million. Diseases like diabetes and Alzheimer's represent a growing threat, not only to patients but to our ability to keep health care affordable. We know that we have to do better in our lifestyles and in our health care system to avoid an enormous disease burden and economic burden on the Boomers themselves, their families, employers, and federal and state governments. The recent research programs funded by EU focus on prevention starting from the younger generations, promoting the adoption of healthier lifestyles, light monitoring of basic bio-signals for early detection of the main pathologies and collaborations/integrations among all the healthcare actors through ICT systems. Also new pharmaceuticals are a vital part of the solution to this rapidly emerging issue.

Therefore the new vision is the *Targeted Approach to Treatment*. Increasingly, the administration of new medicines will be guided by predictive evidence from genetic and other molecular tests. The expectation of "personalized medicine" is that these tests will reveal whether an individual is likely to respond well to a drug, or avoid toxic side effects. A targeted approach to treatment can ensure that each patient receives the right medicine at the right time. About a dozen of such treatments are in use today, and the field is growing rapidly. Since molecular diagnostic tests can reveal a patient's susceptibility to disease, they can also guide preventive treatment before symptoms arise. The emergence of personalized medicine will shift the focus of medical care from "disease treatment" to "health care management."

When speaking of personalized medicine and anywhere anytime healthcare services, this means that for the future of Healthcare, it is necessary to look beyond the Hospital. Nowadays Healthcare services are hospital centered; but this infrastructure is set up to rely on high-cost facilities. If we need—as we strongly need both for economic reasons and, maybe more, for an improved quality of life—to lower costs, this will require innovative treatment at a new type of facility or preferably at home. This means dedicated innovation in new easy-to-use and safe technology for "distributing" the health services to the user-patient point-of-care. In actual fact, technology is also profoundly changing from this standpoint: it is moving from being concentrated at the point of care in the hospital to being in the hands of patients and caregivers. For the last century, technology has been geared toward replacing the dialogue between the patient and the physician; it was designed to reveal the "truth" about medical conditions that most consumers can barely comprehend.

But chronic disease management challenges even this and requires consumers to participate in maintaining and managing their health—often on a day-to-day basis. The smartphone and iPad communication revolution, where technology becomes an enabler of positive social and environmental change, allowed (source Kaiser Permanente) [8] delivering 5 million visits via videoconference (Skype), telephone, and email in last two years in the U.S.A.. Shifting health care from a "point of service" clinical model to an ongoing dialogue between patients and their providers is a profound social and technological shift. This "revolution" is particularly relevant in chronic disease management, e.g. asthma, diabetes, obesity, and also cancer. This issue is rapidly and largely growing with the increase of life expectancy. The elderly are the main target-users of this action but they are also not so expert in technology: this situation will be probably solved with the next "technological" generations, but up to now this gap is a criticism.

In this way we can say that the future of healthcare is the drive for more personalized care, that will increase the quality of the service and will allow to control costs. For many years, healthcare trends have looked unsustainable—ageing populations result in more people needing care, while a declining workforce means fewer people to pay for, and actually deliver, that care. At the same time, an inexorable rise in the costs of care linked to long term conditions reflects the impact of modern lifestyles, as well as positive advances in medicine and public health. Up to now it has been possible to absorb these costs incrementally, but with the consequence that either a significant proportion of the population loses access to care for all but emergency procedures, or the debate focuses on the rationale for controlling access to treatments.

Spending on healthcare is becoming unsustainable. Even with a raft of changes, a recent European Union study raised concerns about the sustainability of public finances. It predicted health and long term costs in the EU would rise on average by about 2.6 percentage points of GDP by 2060, a 33 % increase. This far from modest prediction already takes into account measures such as prolonging working life; increasing participation of youth, women and older workers; reforming tax and benefit systems; and making health and long term care systems more efficient.

Thus, investing in healthcare requires a long term vision, and technology is a key point. The complexity of healthcare has resulted in a relatively late adoption of information technology, yet it is now clear from experience that health systems which are able to use information to drive quality and provide more preventive care can achieve remarkable results. Effective investment in health requires a holistic, long term view of the issues.

If the future of health lies in the use of information to drive for more personalized care in a way that opens up access and drives up quality while controlling costs, then the prospects in healthcare are:

- New governance and new service models;
- Telehealth exploitation, that means to empower people to manage their own health and wellness;
- Health information vision, i.e. transforming healthcare delivery from volume of care to quality of care.

In this scenario, it is clear how investing in new technologies is strategic. It is strategic but also economically challenging, both for the investments and also for their return. In this situation IPR issues can become essential: this is probably the main reason for the incredible increase in patent applications occurred in last years. For example, in the last decade more than 500 patent applications were claimed in the field of wearable sensors [1]. Thus in this book and specifically in the next chapter the methods to protect and exploit innovation will be presented and discussed in detail.

References

1. Andreoni G, Barbieri M, Piccini L (2011) A review of the intellectual property right in the field of wearable sensors and system. Int J Comput Res 18(3/4):269–285
2. Donahue S (2010) Can cloud computing help fix health care. Cloudbook J 1(6)
3. Heads in the cloud. Digitising America's health records could be a huge business. Will it? Mar 31st 2011 | NEW YORK | from the print edition, http://www.economist.com/node/18486153
4. Horowitz BT (2012) Cloud computing in health care to reach $5.4 Billion by 2017. http://www.eweek.com/c/a/Health-Care-IT/Cloud-Computing-in-Health-Care-to-Reach-54-Billion-by-2017-Report-512295/

Web References

http://www.microsoft.com/health/en-us/initiatives/pages/cloud-services-for-health.aspx
http://www.oracle.com/us/industries/health-sciences/solutions/index.html
http://www.cisco.com/web/strategy/docs/gov/fedbiz031611Healthcare.pdf
https://healthy.kaiserpermanente.org/html/kaiser/index.shtml
http://www.epo.org/

Chapter 3
Innovation and Rights

3.1 Introduction

Innovation could be incremental or disruptive. Obviously this second opportunity is more strategic and we can speak about inventions. In general inventions can only be a process and not a product: this distinction is legislative and not phenomenological. In fact, anyone who discovers or invents a new product but does not know how to build it has not achieved a patentable invention.

However, there are patents concerning products and/or processes; if a new way to produce a known product is found then you can patent the process; If the product is new (and inventive) you can patent the product.

An invention does not often consist in the idea of the product but in its destination, in this case the inventive contribution resides in the use of the product (and it is proper to limit the monopoly to use and not to the product itself) [15].

You have an invention of new use when the knowledge (also said as the technical teaching how) to obtain the product was already known in the state of the art, but the particular function was not practicable (for a person skilled in the art) in an obvious way.

There is no a statutory definition of invention.

Article no. 45(2) of the Italian Code of Industrial Property (*Codice della Proprietà Industriale*, CPI) defines how to identify non-patentable inventions, such as the discoveries (e.g. a new substance in nature or a new property of a known material or the mechanism of action of a drug), scientific theories (e.g. the theory of semi-conductivity), mathematical methods or aesthetic creations.

Instead, the solution of a technical problem (e.g. the use of a substance found in nature, or a new semiconductor) is patentable.

The exclusive right provided by the patent system has nothing to do with the expressive phenomenology of the technical solution (product, process), but rather relates to the technical teaching deriving therefrom (which can also be expressed in one or more products or in a method).

G. Andreoni et al., *Developing Biomedical Devices*, PoliMI SpringerBriefs,
DOI: 10.1007/978-3-319-01207-0_3, © The Author(s) 2014

Another differentiation between main inventions and relative or dependent inventions can be identified. The former are those ones reached without a link with other inventions, even if arising from prior knowledge. Instead derived inventions take as direct starting point a previous invention.

Dependent inventions can be divided into the following three types:

- Improvement: starting from a previous invention, inventors define an incremental improvement;
- Transfer: starting from a previous inventive concept, inventors identify a new use that is an original application in a new technical field;
- Combination: by combining earlier inventions or elements from previous inventions, a new technical solution is found.

Obviously, these inventions must be characterized by an inventive step, in order to avoid being an implementation variation that does not produce a different and better effect of the device or process.

From a legal point of view, there is an important difference between independent and dependent inventions. By virtue of legal protection, dependent inventions can be implemented only with the consent of the holder of the previous patent (for example, a new process for manufacturing a product already covered by the patent belonging to others).

The industrial patents are divided into two categories:

- patent for industrial invention;
- patent for industrial design, which is again divided into utility models and industrial designs.

The patent for utility model is distinguished from the patent for invention in that it protects enhancements of existing products, which result in the increased usefulness of the products themselves, rather than in new solutions to technical problems.

A patent for utility model must protect a concrete object, so that it does not apply to processes. Compared to a patent for industrial invention, it has a duration of 10 years from the filing date, divided into two five-year periods, with maintenance fees which can be paid in a lump sum or in two installments, one at the time of filing and the other at the expiration of the first five-year period.

The patent for design is meant to protect the aesthetical features of any industrial or handicraft item, the whole product or a part thereof, as resulting in particular from characteristics of the lines, contours, colors, shape, texture and/or materials of the product itself and/or its ornamentation.

Designs may be registered provided that they are new and show an individual character.

Designs are considered to be identical if their features differ only in trivial details.

The requirement of individual character is met if the impression that the design evokes in an informed user differs from the overall impression produced on such a user by any design which has been disclosed before the filing date of the application for registration or before the priority date which may have been claimed.

The registration of a design has a duration of 5 years from the application filing date, which may be extended for one or more 5-year periods, up to a maximum of 25 years.

A single application may cover the registration of several designs (up to 100), provided that they are implemented or embodied in objects belonging to the same classification [7].

3.2 Types of Industrial Property Rights

3.2.1 Patents

A patent is a legal document granting to the applicant (a company or an individual) a temporary (for 20 years) and territorial (with reference to the country/ies in which the patent has been extended) monopoly.

The ownership of a patent confers the right to prevent third parties from reproducing the invention, i.e. producing, using, marketing, selling or importing the invention in the territory in which the title was filed and granted (except for scientific purposes and for private use).

Therefore it is a negative right: its ownership does not guarantee the holder the corresponding positive right to implement the invention.

To be really implementable, no earlier patents (i.e. exclusive rights granted to others) must exist about the technical solution.

In short, a technology which is new and inventive is patentable; a technology which does not fall into the scope of protection of patents and/or patent applications of others can be freely implemented; a technology which falls into the scope of protection of pending patents is an infringement of the same.

Patentability and feasibility are the two required conditions for the optimal exploitation of the invention [12].

A patent application filing may be both national (by virtue of the Paris Union Convention, this can be extended to other countries within 12 months from the filing date) and foreign. However, foreign filing can only occur after 90 days from the filing of the Italian application, required in order to assess whether the innovation may be of interest to national defense. In fact the territorial extension is possible only under authorization by the Ministry of Defense. On the other hand, a first foreign filing either European or PCT patent application, should obtain the authorization granted by the Ministry of Defense.

3.2.1.1 Structure of the Patent

Patents mainly consist of four sections:

1. first page (bibliographic data and abstract of the invention);
2. description;
3. claims;
4. drawings (optional).

The title is usually written in a very general way (sometimes even too much so: for example, "Method" can be found in databases in as many as 100,000 occurrences).

If sufficiently descriptive, the title, should help in the classification and in the state of the art (*prior art*) search.

Instead the abstract should enable the comprehension of the main features of the invention, but usually simply reports the main claim (and possibly a drawing).

Box 1 Example of abstract (PCT patent application No. WO 2006/106429)

Method for the deposition of steel on at least one surface portion of a nitride metallic element comprising a phase (a) releasing nitrogen from at least one surface portion of said nitride metallic element and a phase (b) depositing at least one layer of molten metal on at least one surface portion of the nitride metallic element.

The description is structurally divided into several sections:

1. state of the art (where the problems encountered in the prior art are emphasized and the technical problem solved by the invention is presented);
2. summary (differing from the abstract appearing on the first page of the patent application) of the advantages of the invention over both scientific and patent prior art;
3. description of the drawings (optional—if drawings are reported in the patent application);
4. detailed description;
5. examples.

The description of the invention must contain all the necessary knowledge so that a person skilled in the art may carry out or reproduce the invention. The information that is part of the common general knowledge of a person skilled in the art is omitted.

The description (and possibly the drawings) are used to allow a better interpretation of the patent, but cannot be used to expand the scope of protection. The description ends with one or more claims.

A claim is a statement that defines the scope of the requested protection or simply a sentence that defines the technical elements that make up the invention (using words or technical terms which are a generalization of the original elements to achieve the broadest possible legal protection).

A claim must, of course, meet the patentability requirements (novelty, inventive step), be clear, concise and supported by the description (which means that the description should be the basis for the claimed subject matter and a claim cannot have a broader scope than what is described, represented in the figures and known from the state of the art).

But, at the same time, a claim must include everything that is an obvious modification or an equivalent (means which perform the same function, basically in the same way to achieve the same result), as well as all embodiments which show the same properties or uses.

A claim can be compared with a fence delimiting the owned land: it should clearly define the border of the owned land (the designed innovative solution), avoiding trespassing on the property of others (previous patents both in force or expired) or public property (non-proprietary, public-domain solutions).

The claims define the scope of patent protection [21]; they can be claims concerning products, processes, pharmaceutical composition, uses.

Box 2. Example of claims (tree structure — PCT patent application No. WO 2006/106429)

1. A method for the deposition of steel on at least one surface portion of a nitride metallic element characterized in that it comprises a phase (a) releasing nitrogen from at least one surface portion of said nitride metallic element and a phase (b) depositing at least one layer of molten metal on at least one surface portion of said nitride metallic element.
2. The method according to claim 1, characterized in that said method comprises repeating said phase (a) various times in order to reduce or in any case greatly limit the presence of nitrogen inside said at least one surface portion of said nitride metallic element.
3. The method according to claim 1 or 2, characterized in that said phase (a) comprises a phase (c) melting said at least one surface portion of said metallic element by means of a heat source.
4. The method according to claim 3, characterized in that said heat source is a laser source susceptible to melting said at least one surface portion of said nitride metallic element.
5. The method according to any of the claims from 1 to 4, characterized in that it comprises a phase (d) processing said at least one steel layer deposited with tool machines.

Box 3. Example of drawings (patent No. IT1364892)

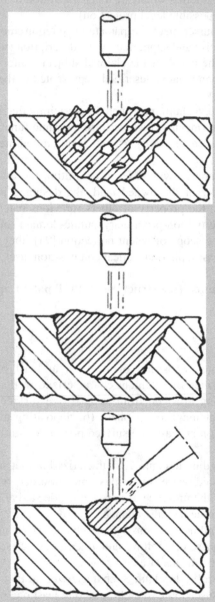

Figures 1a and 1b are elevated views showing the first step of a preferential embodiment of a method for the deposition of steel on a nitride metallic element according to the present invention;

Figure 2 is an elevated view of the second step of a preferential embodiment of a method for the deposition of steel on a nitride metallic element according to the present invention.

It is permitted to narrow the claims by combining, for example, the main claim with one or more secondary ones.

Any amendments should be "limited to claims without incurring the use of elements contained only in the description".

In the claims section (written according to European practice) we can generally distinguish the following elements:

- the preamble describing the prior art;
- the transition part (a sentence of the type *characterized in that, consisting of, comprising*);
- the list of the technical features of the invention.

Claims can be:

- independent, i.e. addressed at the essential characteristics of the invention, except for implied characteristics;
- dependent, i.e. addressed at *particular embodiments*, which include the essential features of the independent claim on which they depend and, possibly, also the additional characteristics of the dependent claims from which they derive. The terms *particular embodiments* also means a more detailed invention.

Generally, a claim is defined by "positive" features: for example, it is written as *a device consisting of X, Y, Z* and not *a device without (feature) K*.

A claim must be presented in terms of "technical features of the invention" (support structure, compound X, steps of an operation: mix X with Y at a temperature of Z, etc.). Statements, for example, concerning commercial benefits or other non-technical features are excluded.

Features other than those entailing "structural" restrictions are admitted: for example, functional characteristics (only if a person skilled in the art can easily retrieve the means to perform that function, without exerting any inventive step).

Claims regarding the use of an invention, meaning its technical application, are also accepted.

3.2.2 The Italian Patenting Procedure

The management of national patents is assigned to the Italian Patent and Trademark Office (*Ufficio Italiano Brevetti e Marchi*, UIBM). It is an administrative office operating within the framework of the Ministry of Economic Development. This office is in charge of receiving national patent applications, and its branch offices delegated as collection centers are the local Chambers of Commerce.

In order to be "receivable", an application must be filed exclusively by these offices or mailed directly to the UIBM. The procedure is described in Fig. 3.1.

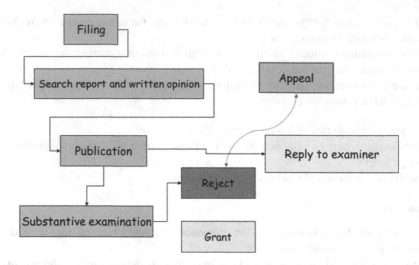

Fig. 3.1 Italian patenting procedure

3.2.3 The Patentability Search Carried Out by the EPO

A Ministerial Decree of 27 June 2008 establishes the rules for carrying out the substantive examination of applications for patents of industrial invention [16].

"The Italian system introduces an examination of novelty but outsources competence thereof".

Indeed, art. 1 specifies that the prior art search is to be carried out by the European Patent Office (EPO), that should receive from the UIBM within 5 months from the filing date the text of the application to be assessed.[1]

The EPO examiner prepares the search report[2]: it may also be partial, if a lack of unity of invention is found. Together with the search report the examiner sends to the UIBM the written opinion which has a "purely informative" value. This operation is due within the following 4 months, then UIBM forwards these documents to the applicant.

Typically the applicant is provided with a valid decision for the later stages within 9 months from the filing date of the patent application. This scheduling allows the applicant to have sufficient information and time for the next decisions about extension of the national application.

---

[1] The application also includes the English translation of the claims [either submitted by the applicant within two months since the filing date (with an extension of one month) or performed with the aid of automatic translators: in this case, the applicant shall have to pay a fee of €200] (Art. 8).

[2] Unless the application relates to non-patentable findings or else the description, the claims and the drawings contain abstruse, inconsistent or contradictory notions [art. 4(4) and art. 5].

UIBM carries out a preliminary assessment and if a patent application shows an *absolutely clear*[3] lack of patentability requirements is directly rejected; in this case the documents are not sent to EPO for evaluation.[4] If the UIBM examiner decides that the patent cannot be granted because requirements are not met, he/she will send a rejection notice[5] *with adequate motivations* to the applicant, who may appeal (art 6) within 60 days and may apply for the conversion of the patent application into a utility model.[6]

After the receipt of the search report (but within the period of publication of the application) and depending on the outcome of the patentability search, the applicant may (art. 5):

- amend (or merge) the claims[7] and/or edit the description[8];
- submit arguments about the relevance of the documents quoted in the search report and clarifications about the eligibility of claims;
- decide to file one or more divisional applications.[9]

After the publication of the patent application, the UIBM should proceed with the substantive examination.

3.2.3.1 Substantive Examination for Italian Patent Applications

The substantive examination (namely the examination of the patentability requirements) is mandatory for all national patent applications filed after 1 July 2008.

The beginning of the examination is notified to the applicant with a written notice.

The claims, the description and the drawings can be amended or even the patent application can be converted into a utility model.

---

[3] Ranieli [1] remarks that "... in the presence of an invention implemented through a computer software, the UIBM may, having analyzed the application for the invention before forwarding it to the EPO, detect the absence of requirements for validity, which however must emerge plainly from the declarations or allegations of the applicant or from known facts. As a matter of fact, such absolute evidence of invalidity can hardly emerge from a summary examination of an application relating to computer-implemented inventions. Unless they relate to the software itself, such inventions always need careful assessment. Accordingly, although possible, it is scarcely likely that the UIBM may deny patent protection without involving the EPO."

[4] This preventive assessment is dual since it is performed both by the UIBM and by the EPO.

[5] The rejection notice is interlocutory and should envisage a deadline for the applicant (pursuant to art. 172.2 of the Italian Code of Industrial Property) to reply to the examiner's inferences.

[6] A well-known industrial property expert maintains that an examination system such as the French one might be more appropriate, since it is a "streamlined system", and he remarks that "... the only patents for invention which the national Office will eventually grant are those cases for which the search report is positive and those cases for which the applicant shall take the trouble to start an examination procedure, including a possible appeal before the Board, to achieve a patent only for Italy. If this were be the case, Italian patent applications would end up by having a mostly "sacrificial" value: they would be filed in order to obtain (at the expense of the UIBM) a search report, only to be dropped thereafter, without even giving rise to the payment of maintenance fees to the UIBM itself."

[7] Without broadening the scope of protection of the patent.

[8] The description and the drawings are meant to construe claims and not to integrate them.

[9] If the lack of unity of invention has been highlighted.

The reply from UIBM examiners is not mandatorily due

However it is necessary to send a reply to the UIBM notice if the search report is negative, so as to avoid the rejection of the patent application.

Instead it is not necessary if the search report is positive: however, office actions are always sent whether the search report is positive or not.

In any case, also when the search report is positive, it is recommendable to send the UIBM a simple notice applying for the grant of the exclusive right title, highlighting that according to the search report and the written opinion the invention is new and not-obvious [5].

3.2.4 The International Patenting Procedures

There are three main types of patent protection at the international level

- national filings;
- the European patent system;
- the PCT procedure.

The first way to extend a patent application abroad is to file in each foreign country of interest, a national application corresponding to the original one.

The applicant of a national patent application can file any foreign applications within 12 months from the national filing date. Every application will follow a proper procedure. Actually, in some countries the patent is granted as a result of a simple administrative procedure, in others the patent is granted after a substantive examination of patentability requirements (e.g. in Germany, Japan, etc...).

A second way to extend abroad a national patent application is the European patent system. It is a procedure that does not provide a single title with validity in all contracting States of the European Patent Convention, but rather a bundle of national patents. The advantages of the European patent system are the following: it simplifies the stages of filing, substantive examination and grant and it allows to obtain individual national patents, all examined in the same way and, consequently, all with the same scope of protection.

The third way refers to the PCT procedure.

The PCT process does not provide the granting of a patent, but a patentability opinion, which is not binding for the national offices: a national filing or a regional stage (EPO, OAPI, ARIPO, etc.) must therefore be carried out in each country of interest.

The PCT procedure allows the applicant to exercise a *filing option* within 30/31 months from the filing date of the priority application in all the Contracting States of the Patent Cooperation Treaty at a relatively low cost.

The PCT process is advantageous when the interests of territorial protection are mainly overseas, when the number of foreign countries where filing is envisaged is particularly high and/or includes States where costs are particularly high (such as Japan).

The PCT procedure is also profitable when the invention is still at an experimental stage and must still be completed or when a license agreement is being negotiated, because it allows delaying the entry into the national phase.

3.2.5 The European Patenting Procedure

The purpose of the European Patent Convention (EPC) was to establish a *uniform procedure* of granting patents for invention (utility models are not envisaged), recognized by all Contracting States (currently 37—see Fig. 3.2).

The headquarters of the European Patent Office (EPO) are located in Munich, The Hague, Berlin and Vienna. European patent applications may be filed either directly at the EPO or at national patent offices (in Italy the UIBM). Patent applications may also be filed online. It is important to point out that the European patent is not an exclusive title, but contains a bundle of national patents.

The European patent granting procedure starts at the *Receiving Section*, which verifies the formal requirements (such as the payment of filing and fees) to establish a filing date.

In failure thereof, the application will be deemed to be withdrawn.

Fig. 3.2 EPC contracting states (EPO)

Once the filing date and number are granted to the patent application, the section that receives applications will classify (partially) and consequently will send the patent application to the competent examining division, which will verify the patentability requirements (novelty, inventive step and industrial application) and prepare the search report to be notified to the applicant.

One of the tasks of the examiner is to define for to the application not only an International Classification (IPC), but also a Cooperative Patent Classification (CPC). The IPC classification of a granted patent may differ from that of the relevant application if the claims have been heavily amended.

On the basis of the claims (and possibly of the description and drawings), the documents which are relevant for novelty and inventive step or which simply are background art are quoted.

In some cases the patent application is written in such a way that the EPO issues an incomplete search report:

- if unity of invention is lacking (unless additional fees have been paid, the search is performed only on the first invention claimed);
- if the invention is not patentable (for example, *business methods*);
- if the invention relates to methods for treating the human or animal body.

In any case a patent application is not rejected at the search stage.

No changes are allowed before the search report.

The request for examination must be made within 6 months from the date on which the publication of the European search report is mentioned in the European Patent Bulletin, upon payment of the examination fee.

The substantive examination is generally performed by the same examiner who conducted the search and prepared the written opinion. Only in exceptional cases the application will be assigned to another examiner, mostly based on the main IPC classication.

If the examination is positive, the applicant should approve the text as granted, file a translation of the claims into the other two official languages and to pay maintenance fees.

In failure thereof, a series of notices from the examiner and replies from the applicant will follow.

Observations from third parties may be also taken into account (art 115 EPC).

If the examining division establishes that reasons exist for the grant of the patent, the applicant shall be invited to file the *declaration of approval* of the text formulation the Office intends to use for the grant of the title.

In lack of assent within the fixed deadline, the application shall be rejected, unless the applicant submits proposals of change of documents; in this event, the procedure shall continue.

The decision to grant the European patent will be effective starting from the day of publication of the notice of allowance in the European Patent Bulletin and will confer the same rights as a national patent (art. 64 EPC).

In the European system the opposition procedure follows the grant and consists of a centralized process of "revocation" and/or "limitation" of the granted claims.

The opposition procedure (see art. 100 EPC for further details) must be written and raised within 9 months from the publication of the notice of allowance; the opposition must contain relevant motivations, and the payment of the opposition fee is requested. Any third party is entitled to raise opposition.

3.2.5.1 Recent Amendments to the European Patent Convention

The main amendments made to the European Patent Convention in force since 13 December 2007 (EPC 2000), are the following [6]:

- a European patent application can be filed in any language[10]: the applicant must file within 2 months a translation into one of the three official languages of the EPO (in failure thereof, the application is deemed to be withdrawn, without possibility to request the so-called *further processing*);
- it is no longer necessary to file the claims, but only the description of the invention or even a simple reference to a former application (a procedure known as *filing by reference*[11]—the applicant shall have to provide within 2 months a certified copy of the former application and a translation thereof);
- the patent application can be integrated by missing parts of the description or of the drawings[12] (within 2 months from the filing of the same or upon request by the EPO), but not of the claims;
- an application filed in any country which is a member of the WTO[13] (the World Trade Organization) can be quoted as priority application and the holder shall have 16 months' time from the date of first filing to claim or amend the priority statement[14];
- the holder can limit or even revoke a granted European patent[15] (this is an *ex parte* proceeding, to be suspended only if an opposition is raised against the grant of the patent);

---

[10] According to the previous Convention, a patent application could be filed mandatorily in one of the three official languages of the EPO: English, French and German, or else in the national language.

[11] The reference to the former application must contain: the filing date and number; the office where it was filed; a hint at the parts included in the filing by reference (description, drawings and claims).

[12] The missing parts must be contained in the priority application.

[13] And not only in a Contracting State of the Paris Union Convention.

[14] The request for the re-establishment of rights must be filed within two months from the natural expiration of the twelve-month period.

[15] There is no time limit to start the procedure; the revocation is effective since the beginning (ab initio) in all the designated countries; in the event of limitation, the Examining Division starts a substantive examination in order to make sure that the amendments actually limit and do not broaden the scope of protection. If the limitation is approved, the patent is published as B3. If the patent has been validated in Italy, the new amended text shall have to be translated in order to be effective in Italy as well.

- the holder can file a *petition for review*[16];
- *further processing* is extended (it covers almost all those cases of non-observance of a time limit) also for the payment of fees;
- art 51(1) has been amended: inventions in all fields of technology are patentable (as established by the TRIPS agreement and by the Strasbourg Convention), although the patenting of computer software is still forbidden:
- Other amendments are:
  - Art. 52(4), now 53(c) (therapeutic, surgical and diagnostic methods are excluded from patentability for ethical reasons and no longer for lack of industrial application);
  - Art. 54(4) (which was removed—now any European patent application filed before but published later is *prior art*, regardless of the shared designated States);
  - Art. 54(5) (relating to the second medical use—instead of the *Swiss-style* claim, protection of the product is allowed);
  - Art. 69 (establishing that the scope of protection is determined by the claims[17]) and the relevant Protocol (a new paragraph has been added,[18] where equivalent elements are mentioned for the first time);
- it is no longer necessary to designate at least one contracting State (all are automatically designated upon filing the patent application);
- the cases where a translation of the priority application need to be filed are limited:
  - (a) in case an opposition procedure is issued,
  - (b) if *filing by reference* was made or (c) if missing parts have been found in the description and/or in the drawings and these are disclosed in the priority document.

Some of these amendments are particularly relevant for examiners: now important elements may be postponed (for example claims, the statement of existence of a priority document) or strategies may be adopted (for example, the limitation of a granted patent) which will certainly affect the examination stage of the patent application and the publication of the same [3, 26].

3.2.6 International Patent — The PCT Procedure

The international PCT[19] patent system allows patent protection to be applied for in 145 member states. The procedure involves the submission of a single application,

---

[16] Only for procedural infringements (for example, the Board of Appeal has not summoned a hearing requested by the appellant) and not for a substantive re-examination of patentability.

[17] The description and the drawings are meant to help interpret the claims, but cannot be used to integrate them and hence broaden the subject matter of the exclusive right.

[18] "For the purpose of determining the extent of protection conferred by a European patent, due account shall be taken of any element which is equivalent to an element specified in the claims."

[19] Patent Cooperation Treaty: a cooperation treaty signed in Washington relating to patents, administered by the World Intellectual Property Organization (WIPO) located in Geneva.

an international search and a subsequent (not binding, optional) examination, followed by grant by the national offices of the designated countries.

The PCT procedure is not a granting procedure, but in practice a procedure to delay the entry into the national phases (where you will have the patent granted, if patentability requirements are met).

A single filing will produce the same effects as a national application; also, the possibility exists to designate regional patents, which means patents with validity in a group of countries.

Currently there are four regional organizations:

- European Patent Office (EPO);
- Eurasian Patent Organization (EAPO);
- African Intellectual Property Organization (OAPI);
- African Regional Intellectual Property Organization (ARIPO).

PCT does not include the protection of industrial designs, trademarks and plant varieties, which fall under other treaties or conventions.

The procedure is divided into three phases (Fig. 3.3):

- filing of an international application;
- the patentability search;
- demand (or preliminary examination).

The minimum requirements for obtaining a filing date are the following:

- at least one of the applicants must be a resident or national of a Contracting State;
- the application must be submitted in one of the languages of the procedure.

In addition, the application must contain at least:

- a statement that a PCT application is requested;
- the designation of States (all from 1 January 2004);
- the name of the applicant;
- a description;
- *at least* one claim.

Concerning the first phase, the application can be filed by any resident or national of a contracting State (an Italian applicant can file an application either at the UIBM or directly at WIPO or EPO offices).

The language used for the filing depends on the receiving Office (English, French, German, Russian, Japanese, Chinese, Arabic and Spanish), provided that it is compatible with that of the competent ISA (*International Searching Authority*).

The national Office performs a formal examination and forwards a copy thereof to WIPO and to the office in charge of the search, which for residents in Italy is the European Patent Office.

The latter drafts a search report (ISR—International Search Report; see Box 4), where any prior art documents which have been retrieved are mentioned, and a written opinion (WO—*Written Opinion*).

Box 4. Example of search report—international PCT patent application No. WO 2006/106429

INTERNATIONAL SEARCH REPORT	International application No
	PCT/IB2006/000984

A. CLASSIFICATION OF SUBJECT MATTER
INV.   C23C4/02   B23K26/34

According to International Patent Classification (IPC) or to both national classification and IPC

B. FIELDS SEARCHED

Minimum documentation searched (classification system followed by classification symbols)
C23C   B23K

Documentation searched other than minimum documentation to the extent that such documents are included in the fields searched

Electronic data base consulted during the international search (name of data base and, where practical, search terms used)

EPO-Internal, PAJ

C. DOCUMENTS CONSIDERED TO BE RELEVANT

Category*	Citation of document, with indication, where appropriate, of the relevant passages	Relevant to claim No.
A	PATENT ABSTRACTS OF JAPAN vol. 013, no. 519 (N-895), 20 November 1989 (1989-11-20) & JP 01 210133 A (FUJI SEISAKUSHO:KK), 23 August 1989 (1989-08-23) abstract	1-5
A	US 4 570 946 A (TSUCHIYA ET AL) 18 February 1986 (1986-02-18) column 2, line 6 - line 46	1-5
A	PATENT ABSTRACTS OF JAPAN vol. 016, no. 287 (M-1271), 25 June 1992 (1992-06-25) & JP 04 075824 A (AICHI STEEL WORKS LTD), 10 March 1992 (1992-03-10) abstract	1-5

☐ Further documents are listed in the continuation of Box C.   ☒ See patent family annex.

* Special categories of cited documents :

"A" document defining the general state of the art which is not considered to be of particular relevance

"E" earlier document but published on or after the international filing date

"L" document which may throw doubts on priority claim(s) or which is cited to establish the publication date of another citation or other special reason (as specified)

"O" document referring to an oral disclosure, use, exhibition or other means

"P" document published prior to the international filing date but later than the priority date claimed

"T" later document published after the international filing date or priority date and not in conflict with the application but cited to understand the principle or theory underlying the invention

"X" document of particular relevance; the claimed invention cannot be considered novel or cannot be considered to involve an inventive step when the document is taken alone

"Y" document of particular relevance; the claimed invention cannot be considered to involve an inventive step when the document is combined with one or more other such documents, such combination being obvious to a person skilled in the art.

"&" document member of the same patent family

Date of the actual completion of the international search	Date of mailing of the international search report
14 September 2006	20/10/2006

| Name and mailing address of the ISA/ European Patent Office, P.B. 5818 Patentlaan 2 NL - 2280 HV Rijswijk Tel. (+31-70) 340-2040, Tx. 31 651 epo nl, Fax: (+31-70) 340-3016 | Authorized officer GONZALEZ-JUNQUERA, J |

Form PCT/ISA/210 (second sheet) (April 2005)

The list of the categories of documents quoted in search reports is presented in Table 3.1.

After receipt of the ISR, the applicant has the right to amend the claims (to limit the scope of protection), before the publication of the application.

Further amendments to the claims, the description and the drawings are only possible if you request an *International Preliminary Examination*.

After 18 months from the priority date, the application is published.

Once the first phase is over, the applicant may choose to request the international examination and start the second phase, or leave the PCT procedure and have the examination performed in the various countries separately.

The exam will require the filing of a *Demand* at the competent IPEA within 3 months from the date of forwarding of the ISR + WO or within 22 months from the priority date. It is optional and not mandatory in the next national stages, but if the exam is positive the European regional phase (for example) more easily results in the grant of the European patent.

In other states a new patentability search may be carried out.

The entry into the national or regional phases is then delayed to within 30/31 months from the priority date.

The national phases consist in the validation of the international patent application in the designated countries and the payment of national fees, the filing of translations and the appointment of local patent agents, as required by the national laws of each chosen country.

The period for entry into the national phase depends on the Designated or Elected Office (if examination has been requested): for the EPO, 31 months are needed.

Table 3.1 Categories of documents cited in search reports

Letter code	Description
X	Particularly relevant documents when taken alone (implies: the claimed invention cannot be considered new or cannot be considered to involve inventive step)
Y	Particularly relevant if combined with another document of the same category
A	Documents defining the general state of the art
O	Documents referring to non-written disclosure
P	Intermediate documents (documents published between the date of filing and the priority date)
T	Documents relating to theory or principle underlying the invention (documents which were published after the filing date and are not in conflict with the application, but were quoted for a better understanding of the invention)
E	Potentially conflicting patent documents, published on or after the filing date of the underlying invention
D	Document already cited in the application
L	Document cited for other reasons (e.g., a document which may produce doubts on a priority claim)

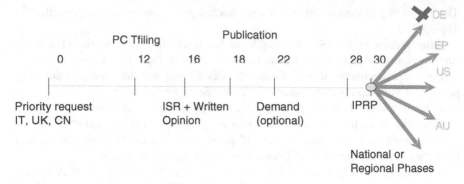

Fig. 3.3 The PCT procedure and its timing (in months)

The PCT system simplifies multinational filing procedures and allows postponing spending. On the other hand, it implies an increase of overall costs. A general suggestion is to apply for a national patent in one's own country and then choose the next procedure among the following:

- apply for a European patent and file separate national applications in non-European countries;
- apply for a PCT patent including the European patent;
- apply for national patents.

3.2.7 Utility Models

If compared to other industrial property rights, there is no harmonized (international or European) legislation on utility models. Only the Paris Convention (1883) hints at them and sets out some general rules, such as national treatment to non-resident holders and temporary protection in international fairs. Therefore, a utility model can be protected on a national basis, but such protection is not provided for by the laws of all states.

Utility models and patents protect different aspects of innovation. A patent for industrial invention refers to a new invention, provided with non-obviousness and liable of industrial application, whereas a utility model consists of an improvement in the object's shape, which does not constitute a solution to a technical problem, but rather confers a particular utility to an already existing product [14].

The improvement obtained by the shape provides the product with an *immediately evident* higher effectiveness and/or ease of use compared to known products, but does not transform the object into something substantially new and different.

A utility model has a shorter duration than a patent: in Italy, its duration is 10 years from the filing date and it is not subject to any examination.

A utility model may also be extended as European patent application.

The system of utility models was originally a German creation, then it has become particularly successful in Asian countries: Japan, Korea, China, Taiwan and Vietnam.

3.2.8 Industrial Designs

About designs, the community rules envisage a dual type of protection: unregistered and registered designs.

Unregistered designs are protected, starting from the date of first disclosure and for a maximum of 3 years, against slavish copying. Registration provides a broader protection (not only from the standpoint of time, namely for 25 years, with renewals every fifth year), namely it confers the following rights:

- it prevents the use not only of an identical design, but also of a design reproducing the same overall impression;
- it excludes third parties from using the design (manufacturing, offer and/or sale, import/export, utilization).

New designs provided with an individual character may be registered (not patented). Assessment is no longer based on the aesthetic value (the requirement of *special ornamentation* was removed). A design is new if, before the application filing date, no identical design or no design differing only in trivial or irrelevant details has been disclosed [19].

The requirement of novelty is absolute: this means it has no limitations in terms of either time or territory. However it is softened by some provisions:

- knowledge by informed users;
- disclosure to third parties under a duty of confidentiality;
- disclosure to third parties entailing abuse against the applicant;
- pre-disclosure by the author during the 12 months preceding the application filing.

A design is provided with individual character if informed users would find it different from other designs available to the public, taking the degrees of freedom of designers into consideration.

The individual character has no aesthetic reference.

Informed users are neither experts or design critics, nor average consumers, but people who is well acquainted with the market of the product where the design is embodied and is able to recognize those aesthetic differences which might escape the attention of occasional purchasers.

Some examples of assessment of the requirements are presented in Annex 1 at the end of this chapter: they show some decisions of the Invalidity Division of the Office for Harmonization in the Internal Market (OHIM).

The individual parts of a complex product are also liable of protection, but only with regard to their outer appearance.

Once embodied in the complex product, the component must remain visible.

Spare parts can be protected only theoretically: in actual fact, the holder cannot prevent independent spare part dealers from reproducing and selling a component whose aim is to repair a complex product and restore the original look.

Law-makers opted for deregulation in the market of spare parts.

The "repair clause" should exclude accessories, because these feature a shape which tends to be free and self-standing, since they are provided with autonomous aesthetic value [11].

A design is also protected by copyright, but only if it is provided with creative character and artistic value.

It is hard to define as well as to assess the above-mentioned requirements: in definition, reference might be made to *the personal imprint supplied by the author to the work*, whereas in assessment reference might be made to *the particular aesthetic value of the work or the strong innovative character compared to current socio-cultural trends*.

> Design [...] as a word is common enough, but it is full of incongruities, has innumerable manifestations, and lacks boundaries that give clarity and definition.
> Design has so many levels of meaning that it is in itself a source of confusion.
> John Heskett, 2002

The design has a more complex nature than invention engineering. Actually, it should be noted how hard it is to define what intent to "industrial design" is and to provide a definition including all the facets of this discipline.

Industrial design is an inseparable mix between art and industry. In particular, the Italian design plays a recognized world-class role which combines elegance and practicality, innovation and modernity.

Industrial design products ensure a precise identity, can increase the value of a product on the market, improving business performance and promoting competitiveness. In addition, the industrial design is an indicator of cultural belonging and identity.

Industrial design protection is an incentive for the development of Italian design in the creation of quality through the use of legal instruments. It also allows authors to obtain exclusive rights by preventing the spread of unauthorized copying and imitation of their project by a third party.

The industrial design—that cannot be defined either as Article or science, or as technology, it has elements from all these three disciplines—does not easily adapt to the models and procedures used for the protection of inventions.

Design is a subject on which, in recent years, the European Commission has either directly or indirectly promoted a number of initiatives, from law-making and regulation of intellectual property rights in the fight against infringement, to the promotion of a number of support activities and European projects to foster research, learning and networking.

Effective solutions to the protection of designs and models should be investigated in view of the peculiarities of this area, if not difficult, design is the definition of the object of protection, it is not easy to provide an exhaustive definition of the concept of invention design, including the various facets of this sector, compared to what you might do instead with inventions of an engineering character.

In this context, the protection of designs has always been held on the background. The case is very often considered "easier" than the utility patent and this modus operandi and related procedures have been inherited.

The filing of an application for industrial design can protect the appearance of any object produced on an industrial scale or handicraft item, such as the characteristics of the lines, contours, colors, shape and surface texture or materials of the product itself or its ornamentation. This definition also includes components that must be assembled into a complex product, or a product formed from multiple components which can be replaced, permitting disassembly and re-assembly of the product.

In particular, the industrial property law distinguishes industrial design from industrial model. The definition of industrial design comprises decorations, shapes, lines, colors, fonts and presentations, while protecting industrial models is reserved for creations of three-dimensional character that give a distinctive look to a product; without the application of said three-dimensional shape to the product, a further technical effect is involved.

3.2.8.1 Italian Procedure

A design is the exterior appearance of a product, or of a part thereof, as evidenced by visible characteristics (shape, texture, contours, materials, ornaments), but protection is not granted to the components of a product visible in the course of its normal use.

Product means any industrial or handicraft item including packaging, graphic symbols and typographic typefaces but excluding computer programs.

It also includes products that are composed of multiple components, which may be disassembled and reassembled to form a complex product, or a product formed from multiple components which can be replaced, permitting disassembly and re-assembly of the product.

Protection is also possible for the components of a complex product as such, provided that they remain visible during normal use of the complex product.

At present there is a gap regarding the components used for re-separation of the complex product so as to restore its original appearance (e.g. car parts); the difficulty of the subject has led the Community legislature to envisage that the Member States are free to maintain or to deregulate the regime of applicable protection. Italy has decided to maintain the principle of non-enforcement of such components (clause must match).

The law also distinguishes protection for:

(a) Industrial Design: Creating two-dimensional character that is attributed to the product and gives it a unique look; this means that it can also include shapes, lines, colors as a decoration.
(b) Industrial Model: Creating three-dimensional character that gives a distinctive look to a product; without the application of said three-dimensional shape of the product, the occurrence of a further technical effect is implied.

The registration of a design is done by submitting an application at the Italian Patent and Trademark Office (UIMB); registration is granted if the design meets the requirements of Novelty and Individual character.

Since design has "peculiar features", some amendments have been introduced so that disclosure is not considered as destructive of novelty if the already existing earlier design could not be reasonably be known (pre-use right) by circles specialized in the field and operating within the European Community.

The pre-use right protects anyone who has used in their own company and kept secret an invention which is later patented by others.

Should an inventor keep secrecy and should another inventor achieve and patent the same invention, the law envisages the right of pre-use, which allows the former inventor to keep on exploiting the invention within the already existing limits of use.

The right of pre-use is granted only if the invention has been used during the 12 months preceding the patent application.

The second fundamental rule establishes that any disclosure of the design before the application filing date or before the priority date prevents the valid registration thereof. Also in this case an amendment was envisaged: a "grace period" has been introduced so that the disclosure made by the author in the 12 months preceding the application filing date is not considered to be prior disclosure.

The grace period is a period of 12 months (6 months for the U.S. and Japan) in which the company can test the product and see if there is feedback from the market. The advantage is that it saves new inventions from "destruction" while giving the company a "little time" to verify the validity of the product on the market. The disadvantage is that if you disclose a model or design in Italy and then apply for protection in other countries outside Europe, where there is a principle of absolute novelty, pre-disclosure prevents from the registration in those countries.

The duration of protection is 5 years, renewable for four additional times, for a total of 25 years. The renewal must be made within 6 months before expiration or after 6 months by applying a surcharge for the additional period.

One single registration can contain several designs, if the products belong to the same class of the International Classification for Industrial Designs. The same registration can protect a set of products with the same characteristics normally sold together, or that have been designed to be used together. In the same application, products considered as variants of two designs can be filed too, but it is necessary that they refer to the same product and are not substantially different from each other.

The registration of a design and industrial model allows for adding the protection of the project with other forms of industrial property. If a new form or a new variant of a shape makes the product/system more comfortable or allows for using the same product in a more effective way, it is possible to apply for a patent or a utility model.

A design is invalid when:

- it does not fullfill the requirements of novelty and individual character;
- it is contrary to public policy or accepted principles of morality;

- it is in conflict with a previous registration;
- if the use constitutes infringement of a mark or an intellectual work protected by copyright;
- it represents flags, signs or emblems of state, official signs or control of inter-governmental organizations.
- it is dictated by technical function.

3.2.8.2 The Protection of Community Designs

In Europe a design can be protected at a national level, by filing the design with a national office in accordance with the specific laws and procedures.[20] But, it can be also protected at a Community level, by virtue of a Community design. Protection allows the applicant to prevent others from using its design without its prior consent, thus encouraging investments in the development of new products.

The most important European directives for the protection of Design are:

- the Directive 98/71/EC of the European Parliament and of the Council of 13 October 1998 on the legal protection of designs.
- the Council Regulation (EC) no. 6/2002 of 12 December 2001 on Community Designs.

Through these laws member States of the EU are encouraged to support the legal protection of designs, to remove the heterogeneity of former national legislations and to clear juridical uncertainties. This last issue is based on the fundamental principle that the protection of designs confers exclusive rights on the shape of a single product, of a complex product or of a particular component, provided that the design is new and has an individual character (Article 1 and 3 of the Directive). The notion of ornament is overcome and any reference to aesthetical features is removed, so that even functional or ergonomic industrial designs may obtain legal protection.

The EC Regulation no. 6/2002 also establishes a new and peculiar instrument for the legal protection for designs, known as Community design, that is a single title with validity throughout the European Union. Meant to coexist with national designs, Community design shares the same basic rules of legal protection with them, so as to guarantee unity of law enforcement regardless of the chosen method or instrument of protection.

Community designs may be either registered or unregistered, but both of them must comply with the two fundamental requirements (shared with national designs) of novelty and individual character.

The procedure to register a Community design is the same throughout the territory of the EU and is available since 1 April 2003; it requires the filing of the special application with the Office for Harmonization in the Internal market (OHIM),

---

[20] For national discipline, reference is made to the relevant paragraph in Sect. 3.7.

located at Alicante; the registration gives the holder exclusive right to implement and market the protected design, either individually or on the products in which it is incorporated, for a period of 5 years, which can be extended to a maximum of 25 years with following renewals.

The unregistered Community design is in force since 6 March 2002; it is acquired by the author free of charge and automatically (without filing any application) starting from the date of disclosure or public use and with a duration of 3 years. The protection granted by unregistered designs is less complete than the one offered by registered designs and covers the design only against copying.

The unregistered design aims at providing legal protection to products having a short trade life, widely represented in the fields of fashion, textile or toys. Although the procedure for registered design is quick, easy and not expensive, it is poorly suitable for products which are often meant to remain on the market for less than 1 year: in that case, the unregistered design offers adequate protection from copying, without needing any cost or formal procedure for its obtainment.

The unregistered design has another important goal: it offers protection against copying to those designs whose registration has been applied for, but which, must undergo an experimentation period in the year after the application filing. During this time, if unregistered Community designs did not exist, these designs might be freely copied without infringements of rights, since they are not yet registered. This so-called "grace period" (involving experimentation for 1 year) is foreseen by the new Community rules and is aimed at avoiding that disclosures of the design before the filing date, made by the author or with the consent of the author, may destroy the novelty of the pending design. In contrast with previous national laws, any disclosure of the design by the author or with the consent of the author in the year following first disclosure does not destroy the novelty of the creation, which can hence be marketed and tested in terms of commercial success, without negatively affecting the validity of a registration applied for within 1 year. During this 1 year experimentation period, the unregistered design protects the product against copying.

3.2.8.3 International Protections

The Hague Agreement is an international registration system that provides the opportunity to obtain protection for industrial designs in the States and intergovernmental organizations that are part of it. The agreement consists of three international treaties: the first was signed in London in 1934, the second in the Hague in 1960 and the last in Geneva in 1999. Currently 48 countries have joined it, including all states of the European Community.

The international registration of the design under this agreement does not permit the grant of a patent design such as the Community design. However, with the international design you can claim priority within 6 months from the date of the first national registration.

Indeed, the international registration is a simplified procedure that allows to submit a single application to the World Intellectual Property Organization by the competent authority of the original country, rather than filing several design applications in the countries of interest.

The phases of examination and registration will be made by the respective patent and trademark offices of each member State in accordance with the applicable laws.

The international registration has an initial term of 5 years and is subjected to national legislation in force in each of the designated States.

The validity of protection may be prolonged for a further period of 5 years up to a maximum of 10–25 years, depending on the internal legislation of each country. In the United States the duration is 14 years, while in Europe, it is 25 years.

The main difference is that the Community design grants one right in the whole territory of the EU, whereas the Hague Agreement gives the possibility to file through one centralized application several national design applications for several design rights.

3.2.8.4 Other Protections

In addition to the registration of the national and/or Community design, designers can protect their products through other legal instruments.

Another form of legal protection is copyright. The law offers each EU Member State the opportunity to rule independently with regard to the requirements for the recognition of copyright to industrial designs.

However, Italy has amended its copyright law by removing the concept of splitting the artistic value from the industrial nature of the product and introducing products with creative character and artistic value as works of industrial design.

3.3 The Patentability Requirements

There are three patentability requirements: novelty, inventive step and industrial application.

The normative references, considering the Italian and the European law, are as follows.

3.3.1 Novelty (Article 54: EPC)

1. An invention shall be considered to be new if it does not form part of the state of the art.
2. The state of the art shall be held to comprise everything made available to the public by means of a written or oral description, by use, or in any other way, before the date of filing of the European patent application.

3. Additionally, the content of European patent applications as filed, the dates of filing of which are prior to the date referred to in paragraph 2 and which were published on or after that date, shall be considered as comprised in the state of the art.
4. Paragraphs 2 and 3 shall not exclude the patentability of any substance or composition, comprised in the state of the art, for use in a method referred to in Article 53 (c), provided that its use for any such method is not comprised in the state of the art.
5. Paragraphs 2 and 3 also shall not exclude the patentability of any substance or composition referred to in Paragraph 4 for any specific use in a method referred to in Article 53 (c), provided that such use is not comprised in the state of the art.

Art. 46 (Legislative Decree No. 30 of 10 February 2005)

1. An invention shall be considered to be new if it does not form part of the state of the art.
2. The state of the art consists of everything made available to the public, in the territory of the State or abroad, before the patent filing date, by means of a written or oral description, by use or in any other way.
3. Additionally, the content of national patent applications or of European or international patent applications designating Italy and being effective therein, as they were filed, the dates of filing of which are prior to the date referred to in paragraph 2 and which were published or made available to the public on or after that date, shall be considered as comprised in the state of the art.
4. The provisions of paragraphs 1, 2 and 3 shall not exclude the patentability of any substance or substance composition already comprised in the state of the art, provided that it is aimed at a new use.

3.3.1.1 Comments

The novelty requirement aims at not allowing for the patenting of what is already known.

The basic concepts are the notion of absolute novelty (the state of the art comprises all the knowledge, wherever it has been made accessible to the public, by any means) and *not accessible to the public*.

The state of the art consists of four categories of concepts [13]:

- the common general knowledge (contained in textbooks and in the main technical papers);
- enhanced knowledge (all patents published by the USPTO, the EPO and the JPO, the patent literature of the most industrialized and of non-industrialized countries specialized in technology);
- the hidden knowledge (graduation thesis, conferences, publications filed in libraries, etc...);
- patent applications filed before but not yet published.

To be novelty destructive, any disclosure must be carried out with regard to skilled persons who can understand and implement the invention.

An invention is not considered to be disclosed if it is communicated to persons legally bound to secrecy either by law (company employees) or by contract (because they have signed a confidentiality agreement).

Summarizing, novelty is lacking when in a single document the reproduction of the invention, or rather of the essential technical features thereof, may be found; then novelty is lost.

3.3.2 Inventive Step (Article 56: EPC)

An invention shall be considered as involving an inventive step if, having regard to the state of the art, it is not obvious to a person skilled in the art. If the state of the art also includes documents within the meaning of Article 54, Paragraph 3, these documents shall not be considered in deciding whether there has been an inventive step.

Art. 48 (Legislative Decree No. 30 of 10 February 2005)

An invention shall be considered as involving an inventive step if, having regard to the state of the art, it is not obvious to a person skilled in the art. If the state of the art also includes documents within the meaning of Article 46, Paragraph 3, these documents shall not be considered in deciding whether there has been an inventive step.

3.3.2.1 Comments

An invention is new if it is different from the prior art and not obvious when it is *substantially* different.

The requirement of inventiveness is intended to exclude the patenting of all that is but a manifestation of the normal technical progress, although it is new.

Compared to novelty, the state of the art that is considered for assessing the inventive step, does not include patent applications not yet published.

An invention lacks inventive step if it reproduces substantially the same function or the same result of a prior art document.

A solution which a person skilled in the art would achieve without a particular intellectual effort, or through a trivial and obvious combination of two or more prior art documents, is not original.

To assess the requirement of inventive step, it is necessary to put a few questions, namely [18]:

1. how can relevant documents be found?
2. how do they combine with each other?

The prior art search cannot be done without limits.

The following knowledge should be excluded:

1. hidden knowledge (because it is not easily accessible to a person skilled in the art);
2. patent applications which are still secret at the date of filing of the application under examination;
3. particularly old prior art, which should not be combined with recent prior art.

Applying the so-called *Problem & Solution Approach* (PSA), only the prior art that solves the same technical problem should be considered.

The PSA is defined as follows:

- identify the *closest prior* art (CPA), typically and ideally consisting of a single document);
- check if there are technical differences between the claimed invention and the closest prior art (if no difference is found, the invention is not new);
- identify the *technical effect* caused by this difference;
- infer the objective technical problem from that difference (so that the difference is the solution to the technical problem);
- assess whether *the objective problem is technical* (if this condition is not met, there is no technical breakthrough over prior art and therefore no inventive step);
- determine whether there are indications in the state of the art that would lead a person skilled in the art to solve this technical problem in the same way as in the invention (if there is any, no inventive step exists).

The closest prior art must be in the same technical field as the invention: the choice falls on the document that describes most technical features in common or most similar functions and/or properties.

After having identified the relevant prior art, the applicant must verify if the documents could be combined to establish the inventive step.

The EPO applies the *could/would approach*: *would* a person skilled in the art achieve the invention or *could* such person skilled in the art implement it?

Already the documents under consideration must contain the teaching to combine the prior art found.

It must be pointed out that the person skilled in the art is an ideal figure: it is a technician, with a general knowledge in the technical field of the invention, with normal working capacity but without inventive skills. In certain cases it is a group of researchers (in multidisciplinary fields such as nanotechnology).

Example

The invention claims:

An alloy having composition <u>ABC</u> for use as a corrosion-resistant material in devices exposed in saline environments.

Document D1 describes an alloy ABD which has good resistance to corrosion (without specifying that it functions in saline environments). The difference is the substitution of C with D in the alloy.

Document D2 discloses a composition suitable for use in the manufacture of boat hulls and comprising B alloyed with C, D or E.

D2 teaches that the elements C and D may be in mixture with AB, whereas C, D and E are elements which may be interchanged with B.

Therefore a person skilled in the art would be motivated to replace C or E for the element D to obtain a good resistance to corrosion in a saline environment.

3.3.3 Industrial Application (Article 57: EPC)

An invention shall be considered as susceptible of industrial application if it can be made or used in any kind of industry, including agriculture.

Art. 49 (Legislative Decree No. 30 of 10 February 2005)

An invention shall be considered as susceptible of industrial application if its subject matter can be made or used in any kind of industry, including agriculture.

3.3.3.1 Comments

The industrial application requirement is suitable to exclude the patentability of inventions which are not technically feasible (e.g. perpetual motion or time machines), which violate physical principles or which were not technically feasible at the time when the patent application was filed.

3.3.3.2 Lawfulness (Art. 53 (a): EPC)

European patents shall not be granted in respect of inventions the commercial exploitation of which would be contrary to *public order* or morality; such exploitation shall not be deemed to be so contrary merely because it is prohibited by law or regulation in some or all of the Contracting States.

Art. 50 (Legislative Decree No. 30 of 10 February 2005)

1. Inventions the implementation of which is contrary to public order or public morality are not patentable.
2. The implementation of an invention cannot be considered to be contrary to public order or public morality merely because it is prohibited by law or regulation.

3.3.3.3 Comments

Since article 50 EPC must be considered restrictively, its application has not been frequent (for example, patenting cherry bombs)

Art. 6 of the Directive on Biotechnology provides some examples of inventions which are not patentable because they are contrary to public morality and public order: processes for cloning human beings, processes for modifing the germ line genetic identity of human beings, uses of human embryos for industrial or commercial purposes, processes for modifing the genetic identity of animals which are likely to cause them suffering without substantial medical benefit to man or animal, and also animals resulting from such processes.

For the conditions of patentability, also the following articles are important other factors.

3.3.3.4 Unity of Invention (Art. 82: EPC)

The European patent application shall relate to one invention only or to a group of inventions so linked as to form a single general inventive concept.

3.3.3.5 Disclosure of the Invention (Art. 83: EPC)

The European patent application shall disclose the invention in a manner sufficiently clear and complete for it to be the carried out by a person skilled in the art.

3.3.3.6 Claims (Article 84: EPC)

The claims shall define the matter for which protection is sought. They shall be clear and concise and be supported by the description.

3.4 Patent Searches

When judging whether an invention meets the patentability requirements (i.e. if it is new and inventive), a deep search in databases (patents, scientific articles, etc…) must be carried out [18].

But before starting the search to retrieve all the relevant documents, it is useful to establish a clear strategy and more precisely it is necessary to:

- identify the essential technical features and not every single detail of the invention (so it will be easier to deduce the keywords for the search query);
- note that the non-technical features are not part of the state of the art and therefore are not to be searched for (non-technical elements are the description of an algorithm, a mathematical method or a business method);
- understand the technical problem solved by the invention (important because to make a technical breakthrough, the invention must relate to a "technical problem" and must be defined in terms of "technical features");

- select the most appropriate database, taking into account the coverage of the data (i.e. the number of documents included and searchable).

Mainly before filing a patent application, it is appropriate to consider the patent information in addition to scientific literature to determine if a technology (or a device or a process) is already known in the state of the art (and therefore avoid reinventing what is already potentially available as knowledge, wasting resources in terms of time and money) or if it has already been patented: if a technology is protected by a valid patent, it cannot be implemented or reproduced without the owner's consent, nor can it patented by others (since novelty is missing).

Other cases in which it is necessary to make a search in patent databases are the following ones:

- before marketing a product (a device or an equipment);
- to monitor competitor activity or a specific technology;
- to establish the guidelines for the research and development activities;
- in oppositions against the grant of a patent.

This information can only be retrieved with a search in patent databases [9].

Sometimes patents are not taken into account because, rather than useful information resource, they are seen only as tools for business or considered too difficult to understand or because access to information is considered too expensive.

However, patents describe details that are not usually found elsewhere.

While scientific papers are widely cited in patents, the opposite is very rare: a recent study showed that only 0.25 % of articles in the life-science industry quote patents [30].

In addition, patent texts are easily accessible (since free databases offer more possibilities to search for information and an increasingly higher number of documents on file) [21].

However, the usefulness of patents as a source of information decreases due to several factors, such as [6]:

- the availability (and thus the completeness) of data;
- the constant increase in the amount of data uploaded in databases (mainly from China, Japan and Korea);
- as a result of globalization, patents filed in all states are important;
- the complexity of the technology.

3.4.1 Tools to Find Patent Information

Different search systems are usable, both free of charge, provided by national or international (EPO) patent offices and by independent producers (Google Patents)] and professional (Derwent, Micropatent, Questel-Orbit, etc…).

The data available in the various databases and the search capabilities may vary considerably, ranging from the simple bibliographic or keyword search (for example, the Patent Abstract of Japan — PAJ-database) to the retrieval of information unavailable elsewhere.[21]

In addition to search, many national offices offer services such as the possibility to store on disk patent texts and information on legal status.

The EPO and the USPTO[22] also provide access to the register that contains data relating to the prosecution of patent applications.

PATENTSCOPE®, the portal of WIPO, allows searching more than 2 million international (PCT) patent applications and graphically displaying the results on the basis of some simple data such as publication date, main inventor, applicant, IPC classification (see Fig. 3.4).

Professional systems offer more possibilities of search and retrieval of data (in terms of both number and type of stored documents and searchable classifications[23]) and especially of statistical analysis of the results. The use of titles and rewritten summaries adds value.

3.4.2 Patent Classification

The classification of patents means to organize, catalogue and index the technical content of such documents so as to be able to identify and search them easily and accurately.

Patent literature is actually so broad that the retrieval of information is hard without the aid of classification.

The examiners of national and international patent offices (the EPO,[24] the USPTO, and so on) are the ones which give one or more classification codes[25] to patent applications.

Although the international patent classification (IPC) is the most widely used, other types of patent classifications exist, for example the Cooperative Patent Classification (CPC), the United States (USPC), the Japanese [(FI—*File Index*) the *F-terms* (*File forming terms*)] classifications, and so on. In any case, all of them have a very large number of categories (see Table 3.2).

---

[21] In the database of the Canadian Intellectual Property Office (CIPO) alone, bibliographic searches on Canadian patents registered since 1869 can be performed.

[22] The USPTO database was the first one to be accessible online since 1994; four years later, the EPO followed the example of the United States with Espacenet.

[23] HGJFH.

[24] EPO examiners classify not only patents, but also technical and scientific papers.

[25] Depending on the complexity of an invention, the number of IPC classes may be considered as the indication of the complexity of an invention and it is related to the number of claims and above all to the number of pages.

Fig. 3.4 Example of statistical analysis with Patentscope®

Classification helps in performing searches (prior art, novelty or patentability) regardless of the used language, because in this way notions or ideas which are sometimes hard to express through words can be searched for; another important help of classification is the availability of abstracts (in English) or even patent texts are not always available (very old patents are classified as images and hence are not searchable through keywords).

Since a classification perfectly matches the inventive notion in no more than 10–20 % of cases, the use of keywords remains at any rate an essential search method.

3.4.2.1 International Patent Classification

The *International Patent Classification* (IPC) is a system to classify and search for patents and scientific articles [28].

Table 3.2   The main types of patent classification

Type of classification	Number of categories
IPC	69,000
CPC	250,000
USPC	430 classes and 140,000 subclasses
FI	180,000
F-terms	350,000

The classification overcomes possible problems related to terminology (the language used in patents is a compromise between the legal and the technical jargons) as well as those arising from the fact that text searching is not always available (for example, the USPTO database contains only images for patents from 1790 to 1976) and finally it is really effective because useful information is not necessarily based on text (for instance chemical formulas, nucleotide sequences or circuit diagrams).

Inventions are classified based on both functional features and possible applications. The IPC divides patentable technologies into eight sections (A–H), in their turn broken down in increasingly detailed levels (classes, subclasses, groups and subgroups).

3.4.2.2   Other Classification Systems

The classification system applied by the European Patent Office (CPC— *Cooperative Patent Classification*) is based on the IPC system, but is more detailed and undergoing to a higher number of revisions.

The CPC scheme is based on the former ECLA subdivisions and includes:

- all former ICO codes;
- almost entirely the former EPO Keywords (KW);
- some entries originating from USPC.

CPC is similar to ECLA but uses longer numbers to replace the old ECLA letters at the end of classes. Now the CPC scheme consists in a total of about 250,000 subdivisions.

The Espacenet database allows users to perform searches in both systems (Fig. 3.5) (IPC and/or CPC).

The Japanese classification *File Index* (FI) is very similar to the Cooperative one (CPC): it is actually a one-dimensional classification system with a hierarchical structure, based on the IPC system but with additional subdivisions (specified by subdivision symbols). In some cases a *file discrimination symbol* in the form of a letter may be added [29].

An example of the FI is reported here: C 01 B 31/02, 101 F

Instead, the so-called *File Forming terms* (F-terms) follows a different standard: it is a two-dimensional matrix system (not deriving from the IPC system) devised

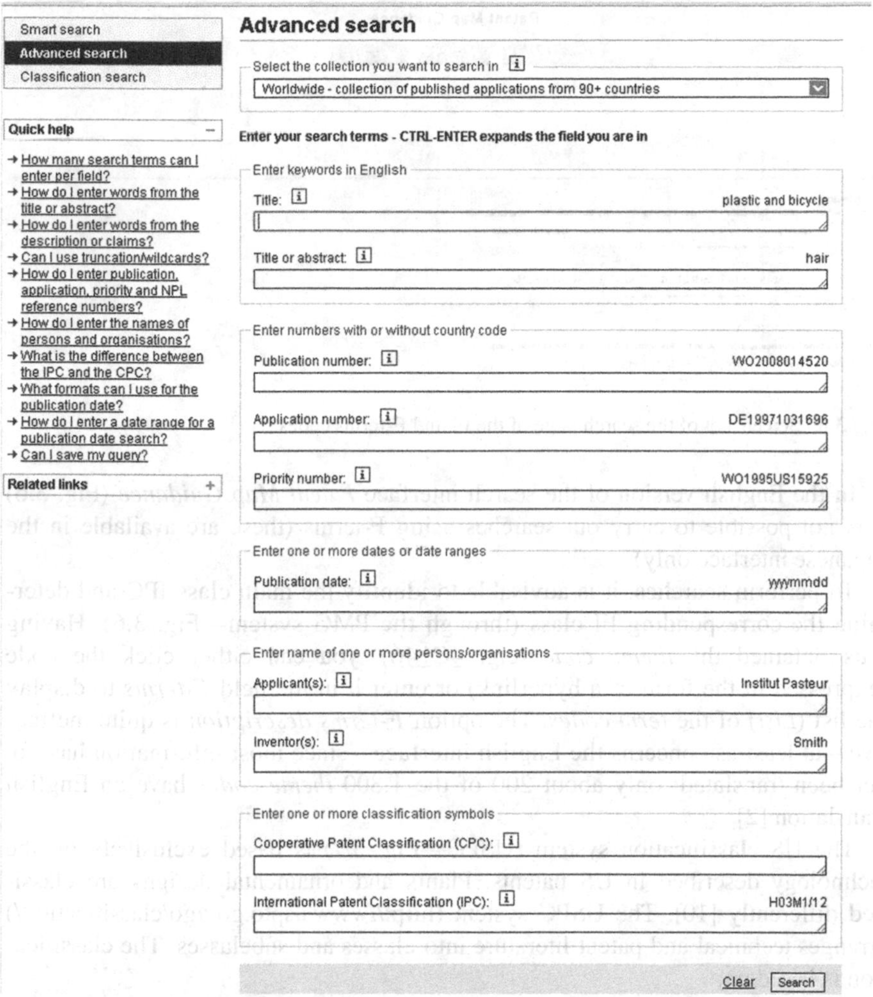

Fig. 3.5 Screenshot of the search page of the Espacenet database

in order to analyze some particular technical fields from differing viewpoints and improve the efficiency of prior searches. F-terms derive from File Index, but in practice they are two distinct and in parallel developed systems.

Each *F-terms* consists of a code (named *theme code*—corresponding to a certain File Index and representing a technological field) and of a code (named *term code*, assigned depending on several technical factors, such as starting materials, purpose, use, products, chemical structure, chemical—physical properties, etc.).

An example of the F-terms is reported here: <u>4C146</u> BC 09

The F-terms system contains much more subdivisions than the CPC system; both systems are organized on a regular basis upon the initiative of examiners and are published in the front pages of only Japanese patent applications.

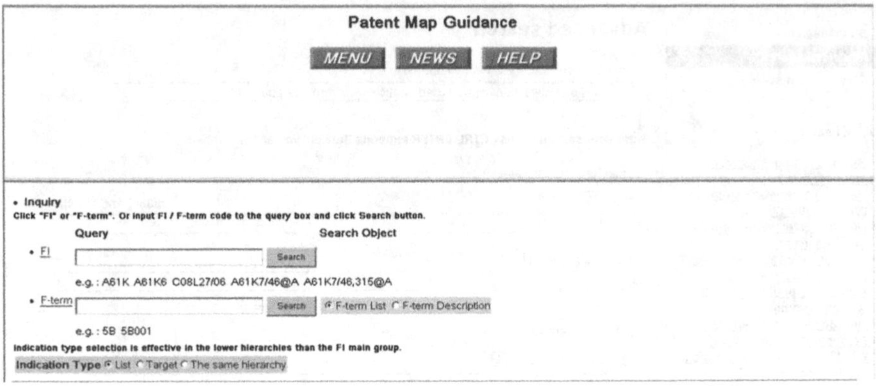

```
                        Patent Map Guidance
                      MENU    NEWS    HELP



• Inquiry
Click "FI" or "F-term". Or input FI / F-term code to the query box and click Search button.
        Query                          Search Object
• FI     [                  ]  Search
        e.g.: A61K A61K6 C08L27/06 A61K7/46@A A61K7/46,315@A
• F-term [                  ]  Search   ⊙ F-term List ○ F-term Description
        e.g.: 5B 5B001
Indication type selection is effective in the lower hierarchies than the FI main group.
Indication Type ⊙ List ○ Target ○ The same hierarchy
```

Fig. 3.6 Screenshot of the search page of the FI and F-terms codes

In the English version of the search interface *Patent Map Guidance* (Fig. 3.6) it is not possible to carry out searches using F-terms (these are available in the Japanese interface only).

To perform searches, it is advisable to identify the main class IPC and determine the corresponding FI class (through the PMG system—Fig. 3.6). Having thus obtained the *theme code* (e.g. 4C146), you can either click the code (expressed in the form of a hyperlink) or enter it in the field *F-terms* to display the list (*List*) of the *term codes*. The option *F-terms description* is quite ineffective—at least as concerns the English interface—since most information has not yet been translated: only about 200 of the 1,800 *theme codes* have an English translation [2].

The US classification system (USPC—Fig. 3.7) is based exclusively on the technology described in US patents. Plants and ornamental designs are classified differently [10]. The USPC system (http://www.uspto.gov/go/classification/) arranges technical and patent literature into classes and subclasses. The classification is based on:

• industrial field;
• utility (a broad notion relating to the function, the effect or the achieved product);
• structure (chemical/physical composition or configuration).

3.4.2.3 Strategies

No single strategy exists to search for a classification code assigned to an invention. For example, based on the IPC system, it is possible:

• to perform an initial search based on keywords in the title and/or in the abstract and then perform a statistical analysis of the codes ascribed to the documents which are deemed relevant; or else

Fig. 3.7 US patent office site and page for searches

- to use the *Catchwords Index* (Fig. 3.8) available in the WIPO website (http://web2.wipo.int/ipcpub/#&notion=cw) (although the number of terms contained therein is limited); or else
- to perform a search with the IPCCAT tool (https://www3.wipo.int/ipccat/): better results compared to the Catchwords Index are achieved (Fig. 3.9).

3.4.2.4 Limits

One of the main difficulties in using classification arises from the circumstance that no unified scheme exists and accordingly the examiners or patent searchers who query several databases must learn to recognize the various classification systems (IPC, CPC, USPC, DEKLA, FI, F-terms, etc.). Some classification codes apply to national patents only (USPC, FI, F-terms) and hence it is preferable to use the IPC and CPC systems.

The primary aim of a classification system is the creation of an effective search tool, but if a subgroup contains a high number of documents, it is unuseful at all! This is the reason why classification codes are subjected to a revision by examiners in the event of new technologies are available or when the size of subgroups is too wide: the purpose is to make patent searches more efficient.

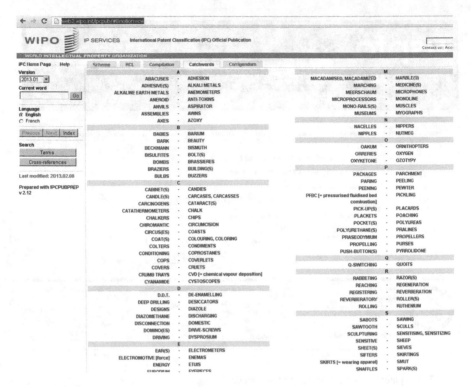

Fig. 3.8 An example of catchwords index

3.4.3 How to Perform a Patent Search

The purpose of a prior art search is to find whether the essential technical features of an invention are already described in some other document (patent or scientific paper); the examiner needs to obtain not the match of words (that is, whether all the keywords used are found in a document) but of concepts, which are better expressed through codes or classification symbols.[26]

Patent searches can be performed either by a series of terms that characterize the invention under examination or by classification symbols, or a combination of both the above-mentioned categories.

Usually a search is the result of the combination of independent mechanisms (text and classification or text and chemical formulas) [1].

The most intuitive way to conduct a search is certainly the use of a set of keywords.

But first the examiner needs to understand the invention in all its aspects by identifying the essential technical characteristics to look for, the elements that represent the core of the invention [26].

---

[26] Classification reduces dependence on the language of the original document.

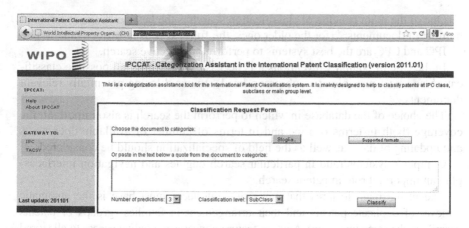

Fig. 3.9 Screenshot of the WIPO web page to search for IPC codes by keywords

Many searches are not successful because they wrongly identify the technical content of the invention and the choice of search terms.[27]

The search carried out only by keywords may be not accurate and precise because it is limited by:

- terminology issues: a patent is a document that is meant to be defended in court (and not to be easily found in a database!) and therefore the language in use is a compromise between the technical and the legal jargons (sometimes the inventor uses neologisms);
- synonyms in different languages;
- incomplete databases (not all databases allow a *full-text* search) and databases that store incorrect data (for example, metadata about the applicants and the inventors);
- specific pieces of information may not be present in the text, but only in drawings or in (mathematical or chemical) formulas, in genetic sequences, etc.

The use of an inadequate terminology can be an obstacle to the achievement of an optimal search (especially in emerging technical fields). A good test to verify that the terminology in use is appropriate is the *International Patent Classification* (IPC).[28]

If most of the documents obtained by a search are classified within an irrelevant IPC section (or class), it means that the correct terminology was not used and it is necessary to define a new set of keywords [25].

---

[27] The percentage of information retrieval varies depending on the synonyms used.

[28] Why was the Dewey decimal classification (conceived in 1873 and used in libraries all over the world) not used and why had a special classification for patents to be devised? Mainly for two reasons: the Dewey classification encompasses all knowledge (but arts, literature and philosophy are not patentable) and it is not sufficiently broken down.

One of the main benefits of classification search is the best coverage in terms of number of documents[29] (for the older ones, the full text is usually not available).

IPC and CPC are the best systems to perform an effective search.

In the case of complex inventions it is necessary to identify all possible classification codes (if possible at a sub-group level), to avoid losing potentially relevant documents.

The choice of the database in which to perform the search is also important: the coverage (both in terms of time and in terms of number of available documents and updating of data) as well as the field of specialization should be considered.

Computer systems (and in particular search engines and navigation interfaces) play an important role in patent search.

One of the main features that navigation interfaces must have is the automatic removal of duplicate patents and their arrangement in families (groups of patents describing the same invention). Another feature concerns the online access to all stored data (text, images, and so on) and to each patent sections (description, claims,...).

An example of search strategy[30] is the following one [24]:

- perform a preliminary search (using keywords) in the *Title* or *Title and Abstract* fields to obtain a minimum set of documents;
- check whether resulting patents are effectively related to the inventive concepts to be analyzed;
- if so, consider the classifications of relevant patents to broaden your search; otherwise, repeat the search with other keywords.

A general suggestion is not to perform a preliminary search in the full text (which might generate a high number of results), unless no results are achieved with the search in the *Title and Abstract* [31].

Some final remarks about patent searches are the following ones. The main difficulties in a patent search can be attributed mainly to the following circumstances:

- the number of patent applications is constantly increasing (especially in biotechnology, organic chemistry and ICT);
- the quality of the abstracts[31] prepared by patent attorneys is quite low compared to that offered by some database suppliers (e.g. Derwent);
- in many cases it is not sufficient to have a searchable text, because it is essential to process images, drawings and formulas;
- databases are incomplete.

To overcome these difficulties it is necessary on one side to act on information technology (development of search engines and navigation interfaces, and the

---

[29] For example, the bibliographic and keyword search on PAJ is limited to patent applications filed since 1976, whereas the FI classification allows searches to be made not only for patents of invention since 1885, but also for utility models (which are not envisaged in PAJ).

[30] Based on the achievement of a set of highly precise data and the subsequent expansion of results.

[31] Among all the sections of a patent, the abstract is the most neglected. The abstract usually quotes the first claim and hence is aimed at protecting the invention rather than informing the public.

availability of more accurate and reliable automatic translators) and on the other side to make more documents available as *full-text* (searchable text) [32].

In addition it is expected that semantic analysis might achieve acceptable levels of accuracy in the very next future.

3.5 Inventions by University Researchers

Pursuant to art. 65 of the Italian Code of Industrial Property (*Codice della Proprietà Industriale*, CPI), university researchers or researchers working in any public research institute are the exclusive holders of the rights arising from their invention, at least as regards the so-called *free research*.

Researchers can then assign their property rights arising from the invention to any third party (including the university).

If there are several inventors, the rights belong to all in equal shares, unless otherwise agreed.

Only those who have actually performed an inventive activity can be acknowledged as co-authors.

Art. 65 applies to researchers of private and state universities alike, but not to employees of private-law organizations such as foundations, even if controlled by public institutions.

Compared to employees of corporations or private organizations (or even public institutions not having research purposes), university researchers may freely use the outcome of their research.

If, after 5 years (maybe a too long a time, considering the obsolescence rate of technology!) from the date of grant of the patent, the researchers has not started the industrial exploitation of the invention (marketing or licensing), the university shall acquire, free of charge, the non-exclusive right to exploit the invention and the property rights related thereto.

The researcher at any rate remains the holder of the rights and might grant other licenses (also for valuable consideration).

In the event of commercial exploitation of the invention, the researcher shall be entitled to no less than 50 % of the royalties, before the deduction of financial charges (art. 65 paragraph 3).

If the research is financed by a private organization or by a public institution differing from the one the researcher belongs to, the rights on the invention belong to the university or to the public research institution (art. 65 paragraph 5) and the researcher only performs professional consultancy [4].

3.6 Technology Transfer Agreements

After an industrial property title has been filed or registered, the problem is how to exploit the invention.

For a company the options are mainly two: (a) implementation by the company itself, and (b) sale, licensing or cross-licensing entrusted to third parties.

Instead university has by contrast a rather restricted scope of action, since it is not allowed to market inventions directly (except through a spin-off) and must therefore identify a company on the market to negotiate a license agreement (either exclusive[32] or non-exclusive) for the use of the patent or an assignment agreement, which however is an outright transfer [23].

A license agreement, which may relate to either patents or know-how or both, is an atypical (free-form) agreement and has some limitations such as [17]:

1. the territorial scope: outside the borders of the States in which the exclusive right title has been validated, the holder has no exclusivity (the invention can be freely implemented);
2. the time scope: the duration of protection is limited (20 years from the filing of the patent application) and affects the duration of the agreement;
3. the scope resulting from the right of use: scope of coverage determined by the contents of the claims;
4. the market sector or field of use.

During negotiations, the potential licensee must check carefully:

- the ownership of industrial property rights;
- the legal status;
- the validity of the titles;
- the time to market;
- the estimation of the fields of use and of the markets;
- the freedom of implementation.

The agreement may provide for:

- a license on later improvements;
- collaboration to ease the implementation of the invention and to obtain the exclusive right titles, since these activities may extend beyond the time when the transfer operation is accomplished, when the object of the agreement is a patent application;
- any sub-licenses.

The costs of maintenance of industrial property rights are usually borne by the licensee, whereas a decision about who will be in charge of management will have to be made.

Other useful tools during negotiations are: confidentiality agreements, letters of intent, joint development agreements.

The only guarantee that the licensor takes on is the one relating to ownership, whereas guarantees are expressly excluded from agreements with regard to:

- the validity of industrial property titles;
- the patentability of the inventions;
- the presence of infringements.

---

[32] The licensee will prefer an exclusive license if additional investments are required before launching on the market.

Consideration is generally in monetary form (reimbursement for the achievement of the titles, assumption of future maintenance costs, payment of a fixed sum, possibly divided into several installments—*down payment* or *lump sum*-, *royalties*, usually calculated as a percentage of sales turnover, or periodic fees) but it can also be non-monetary (*cross-licensing*, contributions in kind—e.g., equipment, research contracts).

The licensee is entitled to act in case of infringement (even if the agreement has not been recorded).

In the event of bankruptcy of the licensee, the license agreement is not automatically terminated, except in the event of winding-up of the company.

If the patent is declared invalid (due to lack of patentability requirements), the license agreement as well is void (Art. 77 c.p.i) and the licensee is no longer required to pay future fees.

Invalidity is not retroactive and therefore the licensor is not bound to return any fees which the licensee may have already paid based on the exclusive rights existing before the declaration of invalidity.

3.7 Examples of Industrial Design Requirements Evaluations

Example 1

Decision of the Invalidity Division of 31/03/2008 - ICD000003200

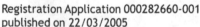

Registration Application 000282660-001
published on 22/03/2005

Earlier design marketed since 2004

The two designs differ in the following features:
- The applicant's design is provided with a hole eligible to contain a connection system
- The function symbols appearing on the keys and controls are different

These differences are regarded as insignificant

Example 2

Decision of the Invalidity Division of 21/01/2008 – ICD000003929

http://oami.europa.eu/bulletin/rcd/2005/2005_061/000334297_0002.htm

Registration Application 0000334297-
0002 filed on 02/05/2007

Earlier design published in December
2004 in the magazine "Asian Sources –
Computer Products"

The two designs are perfectly identical in shape and color

Example 3

Decision of the Invalidity Division of 31/08/2006 – ICD000002640

http://oami.europa.eu/bulletin/rcd/2005/2005_016/000273693_0001.htm

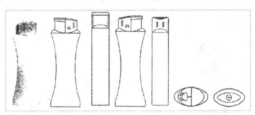

Registration Application
000273693-0001 filed on
29/12/2004

Earlier design No. CN3362229D filed on
07/04/2004

The Community design integrally reproduces the basic features of the earlier
design. The presence of horizontal and vertical lines appearing on the Community
design is considered irrelevant.

Example 4

Decision of the Invalidity Division of 31/03/2008 – ICD000004166

http://oami.europa.eu/bulletin/rcd/2006/2006_104/000574462_0001.htm

Registration application 000574462-0001 filed on 09/08/2006

Earlier design marketed since March 2005

The differences found in this case with regard to the belt and to the sole surface are not consideredas irrelevant details; no identity exists between the two designs. Such elements are not sufficient to convey to informed users an overall impression differing from the one of the earlier design.

Example 5

Decision of the Invalidity Division of 21/06/2007 – ICD0000002848

http://oami.europa.eu/bulletin/rcd/2005/2005_050/000316310_0001.htm

Registration application 000316310-0001 filed on 31/03/2005

Earlier design No. 000241609-0001 published on 25/01/2005

There is no identity between the two designs: the differences (element resembling a diode-shaped lamp bulb and insect-shaped decoration) were not considered irrelevant details. A comparison between the two designs conveys an overall impression of similarity: design No. 000316310-0001 is not provided with individual character.

Example 6

Decision of the Invalidity Division of 24/07/2007 - ICD0000003291

Registration application 000202155-   Earlier design marketed since 1947
0006 filed on 15/07/2004

A difference is found between the two designs: the horizontal seat of the earlier
design is rush-bottomed, whereas the one of the registered design is made of
smooth, unpatterned fabric. This difference is plain to see and cannot be dismissed
as irrelevant details. These differences were not considered as sufficient to convey
to informed users a different overall impression.

References

1. Adams S (2003) Patent searching without words—why do it, how to do it?. http://www.freep
   int.com/issues/060203.htm#feature
2. Adams S (2008) English-language support tools for the use of Japanese F-term patent subject
   searching online. World Patent Inf 30:5–20
3. EPC 2000 and its impact for patent searchers (2007) Patent Information News 1, pp 1–2
4. Bax A (2008) Le invenzioni dei ricercatori universitari: la normativa italiana. Il Diritto
   Industriale 3:205–212
5. Bosotti L (2009) Examination: Does It Matter? Notiziario dell'Ordine dei Consulenti
   in Proprietà Industriale. Newsletter of the Association of Industrial Property Attorneys,
   pp 13–15
6. Bosotti L, Mauro M, Modiano M (2008) EPC2000 Alcuni interessanti commenti. Notiziario
   dell'Ordine dei Consulenti in Proprietà Industriale (Newsletter of the Association of
   Industrial Property Attorneys), pp 14–21
7. Caraher V (2008) The evolution of the patent information world over the next 10 years: A
   Thomson Scientific perspective. World Patent Inf 30:150–152
8. Dragotti G (2001) Appunti di diritto industriale. Available at http://www.dragotti.it/images/
   dispense.pdf
9. Falasco L (2002) United States patent classification: system organization. World Patent Inf
   24:111–117
10. Falasco L (2004) Bases of the United States Patent Classification. World Patent Inf 24:31–33
11. Francetti M (2007) La protezione come disegni e modelli delle parti staccate dei prodotti
    complessi e la disciplina dei pezzi di ricambio. In: Petraz G (ed) La protezione della forma.
    Giuffrè Editore, pp 215–224
12. Franzosi M (1996) La determinazione dell'ambito di protezione del brevetto. Il Diritto
    Industriale, Part I, pp 21–23

13. Franzosi M (2001) Novità e non ovvietà—lo stato della tecnica. Rivista di Diritto Industriale, Part I, pp 63–79
14. Franzosi M (2008) Invenzione e modello d'utilità. La Convenzione di Monaco comporta il rifiuto della distinzione qualitativa. Rivista di Diritto Industriale, Part I, pp 159–170
15. Franzosi M (2008) Definizione di invenzione brevettabile. Rivista di Diritto Industriale, Part I, pp 18–35
16. Giudici S (2009) Note al D. M. 27 giugno 2008. Rivista di Diritto Industriale, Part III, pp 8–10
17. Granieri M (2010) La gestione della proprietà intellettuale nella ricerca universitaria. Invenzioni accademiche e trasferimento tecnologico. Il Mulino
18. Hassler V (2005) Electronic patent information: an overview and research issues. In: Proceedings of the 2005 symposium on applications and the internet workshops (SAINT-W'05)
19. Jandoli V (2003) L'esame dell'altezza inventiva. Rivista di Diritto Industriale, Part I, pp 280–292
20. Lavagnini S (2004) I requisiti di proteggibilità del design. In: Studi di Diritto Industriale in onore di Adriano Vanzetti. Giuffrè Editore, pp 132–141
21. List J (2008) Free patent databases come of age. World Patent Inf 30:185–186
22. Liuzzo L (2007) L'interpretazione del brevetto dopo EPC 2000. Notiziario dell'Ordine dei Consulenti in Proprietà Industriale (Newsletter of the Association of Industrial Property Attorneys), pp 6–7
23. Mastrelia D (2010) Gli accordi di trasferimento di tecnologia. Giappichelli Editore
24. Michel J (2006) Considerations, challenges and methodologies for implementing best practices in patent office and like patent information departments. World Patent Inf 28:132–135
25. Nijhof E (2007) Subject analysis and search strategies—has the searcher become the bottleneck in the search process? World Patent Inf 29:20–25
26. O'Keeffe M (2008) Manifesto for better patent searches and more economical patent prosecution. World Patent Inf 30:1–3
27. Pallini D (2007) EPC2000: una breve guida ai cambiamenti. Notiziario dell'Ordine dei Consulenti in Proprietà Industriale (Newsletter of the Association of Industrial Property Attorneys), pp. 13–17
28. Rampelmann J (1999) Classification and the future of the IPC—the EPO view. World Patent Inf 21:183–190
29. Schellner I (2002) Japanese file index classification and F-terms. World Patent Inf 24:197–201
30. Schwander P (2000) An evaluation of patent searching resources: comparing the professional and free on-line databases. World Patent Inf 22:147–165
31. Seeber F (2007) Patent searches as a complement to literature searches in the life searches—a 'how-to' tutorial. Nat Protoc 2(10):2418–2428
32. Starešinič M et al (2009) Patent informatics—the issue of relevance in full-text patent document searches. Online Inf Rev 33(1):157–172

Web references

http://www.epo.org (website of the European Patent Office—EPO)
http://www.uspto.gov (website of the United States Patent and Trademark Office—USPTO)
http://www.uibm.gov.it/ (website of the Italian Patent and Trademark Office—UIBM)
http://worldwide.espacenet.com/ (website of Espacenet)
http://patft.uspto.gov/ (website to search for US applications and patents)
http://www.wipo.int/patentscope/search/en/search.jsf (website to search for PCT patent applications)

http://www.uibm.gov.it/uibm/dati/default.aspx (website to search for Italian national applications, patents, designs and trademarks)

http://brevets-patents.ic.gc.ca/opic-cipo/cpd/eng/introduction.html (website to search for Canadian applications/patents)

http://eng.kipris.or.kr/eng/main/main_eng.jsp (website to search for Korean applications, patents, designs and trademarks)

http://www.ipdl.inpit.go.jp/homepg_e.ipdl (website to search for Japanese applications, patents, designs and trademarks)

http://english.sipo.gov.cn/ (website to search for Chinese applications, patents, designs and trademarks)

Chapter 4
Case Study: Wearable Systems

4.1 Introduction and Reference Scenario

All people consider health the main and essential value for quality of life, both at a personal and social level. Thus society dedicates a lot of resources to the development of technologies and their impact in the medical field is enormous.

Now, together with the concept of evidence-based medicine, we can also talk about technology-based medicine in several applications. Technology also allows for the development of the concept of quality of care: it is no longer enough to have diagnosis and advanced therapies but well-being, pleasure and comfort have become indispensable too [1].

In the modern vision of Healthcare there is also a trend to user participation in prevention and treatment. This refers to active collaboration of users and their families in the process of basic health (prevention, diagnosis and treatment) through the use of new technologies such as the monitoring of physiological parameters. The future of health care is based on new devices that can provide more accurate and personalized diagnosis and treatment, accessible anywhere, anytime.

Shorter hospitalizations and better home care are the objectives for the current organization of national health services. This can be achieved only through technologies aimed at monitoring patients anytime and anywhere for disease prevention, follow-up and support for rehabilitation that can be integrated with the mobile communication network that is already widespread.

In this context, wearable sensors represent a proposal of strategic interest.

Initially, the interest in wearable systems for health arose from the need to ensure and extend health services out of the hospital and monitor the state of health for a sufficiently long period of time.

Wearable Biomedical Systems (WBS) can be defined as integrated platforms applicable on/to the body to provide "medical activity": from monitoring biomedical parameters to drugs delivery and/or supporting/replacing lost functions (assistive technologies) [2].

G. Andreoni et al., *Developing Biomedical Devices*, PoliMI SpringerBriefs,
DOI: 10.1007/978-3-319-01207-0_4, © The Author(s) 2014

WBS are complex systems that embed several technological innovations (in electronics, material, processing and algorithms, materials, ...), but above all they address an important issue in the management of patients: the acceptance and use of WBS thanks to the user-friendly approach that allows doctors to collect important information and clinical data in the daily lives of the subjects. Indeed, simple direct observations on the impact of clinical interventions on mobility, the level of autonomy, and quality of life can be made by means of wearable systems with minimal intrusiveness or significant impact on the real lives of patients. These factors are crucial, for instance, in chronic pathology management.

This is made possible thanks to the development and evolution of smart garments, i.e. clothing that can become solutions to continuously monitor patients through the non-invasive measurement of biomedical parameters, by integrating directly in their fibers a technological architecture of miniaturized sensors and processing systems. Such sensing elements, incorporated into the fabric of the garments, have opened endless possibilities for monitoring patients everywhere and for long periods of time. This is of particular importance in modern medicine and the future, with citizens-patients becoming active players in the processes connected to their health (prevention, monitoring, treatment). Being wearable and "normal" for a subject (looking like a standard underwear item), WBS represent an ideal platform for non-intrusive, continuous, remote multi-parametric monitoring of the health status for primary and secondary prevention, early detection and management of diseases (e.g. cardiovascular and/or respiratory pathologies), but also as a support for the care of the elderly or disabled people [3].

Generally a WBS is composed of four main elements [4].

The first component is the first Human–Machine-Interface layer constituted by a transducer able to sense bio-electrical/chemical/physical properties or signals from the human body, or otherwise to deliver to the body itself electrical (e.g. small currents or voltages or electro-magnetic fields), or chemical drugs or other substances, or physical signals/actions (e.g. mechanical vibrations, heating, etc.). For these purposes different solutions are applied: from woven/knitted/embroidered textile electrodes for sensing, to micro/mini actuators for mechanical actions (such as vibrations at different frequencies) or chemical delivery of substances.

The second part is the connection solution, that is the input way to the case of the WBS where electronics is inserted. In its simplicity this is a crucial point, up-to-now poorly studied in WBS, but solutions are already available among industrial mini-connectors. The ongoing systems' miniaturizations will push this research towards new solutions.

The third component is the case that is directly related both to the electronics that it has to contain, and to the body location where it is placed onto, and to the body support or fixing element to the body itself.

About this issue, to define body positions, case dimensions, and fixing/support elements, some interesting studies and experiences were carried out by Gemperle et al. a decade ago. Instead, new technological trends push solutions towards textile integration and flexible electronics.

In fact, the next component of the system is electronics, which can be divided into the analogue front-end for signal sensing and pre-processing, or for the conditioning circuits of the micro-actuators, and the digital part for control and data pre-processing.

Finally the last element is related to output signal/information and this is done through two integrated layers: the first one is the second physical Human–Machine-Interface, i.e. buttons, display, LEDs, icons and any other kind of interface commands; the second layer is related in particular to the communication facilities which the electronics is equipped with, which include the firmware to pre-process data and the communication transceiver, where several options could be adopted: from mobile GSM/UMTS device, to satellite modules, from standard or dedicated ISM radio chipset, to commercial Bluetooth/WiFi embedded or integrated radio, up to ZigBee (particularly interesting to build up distributed body sensors networks) or Wireless USB options.

Therefore it is easy to understand that for continuity of care in a social-oriented and no more hospital-centered service, WBS can become strategic and offer clinical tools such as market development. Hence the research and industrial interest in WBS is huge.

This opened an enormous rush to strategic IPR in this field. In the last decade, about 600 patent applications were issued worldwide [4].

4.2   Research in Wearable Biomedical Systems

Human health is one of the issues that has always driven research and innovation, but this process is often done in single sectors, and only after some time and efforts the real integration between disciplines was started [1]. The same biomedical technologies are a paradigmatic example and in these ones it is particularly true that there are devices with high technological content but whose usability is often critical even for very experienced users and poorly studied from the point of view of the comfort of the patient. This means to adopt the most common technology-driven approach for innovation.

Today, ergonomics in healthcare cannot be delayed any longer: high quality new devices are needed to provide care and a safe and effective treatment to patients, as well as to ensure comfort, health and safety for any user (clinical or not, such as relatives).

Thus we must turn to a user-driven innovation approach: it is mandatory (and also the law is going to introduce this requirement for the quality certification of new devices in the next two years) to correctly identify user needs and translate them/incorporate them into the design of the product/environment/interface/service as an essential element of this success. In fact, these health systems are almost always used by many different actors, such as health professionals, physicians, patients and their families. This leads to great difficulties in the design, because it is necessary to consider many different points of view, not to mention the needs of different users. Ergonomics represents an opportunity to encourage a more general consideration of the user, of his/her abilities, mental and physical limitations and requirements during development.

The human-centered approach for the development of methods, products, environments, systems and services in the field of healthcare requires also integration of knowledge: architects, engineers and designers work closely for the study, analysis, evaluation and design of new solutions. Multidisciplinarity and technical and cultural "contamination" and cooperation are the key words that lead to innovation across its pillars: human, technology and design. It is needed to create a new complex and multifactorial process in which technological factors, organizational and human dimensions must find a balanced mix for a full success of the outcomes.

Wearable technologies in general and WBS in the healthcare sector represent a new field developing according to these assumptions.

The history of the WBS is relatively recent: the first studies date back to 1996 in the United States (Fig. 4.1).

Initially, research in wearable systems for health was oriented to technological integration of existing platforms: thanks to the microelectronics evolution, the research focus was the miniaturization and the integration of biomedical technologies to develop new compact systems to measure physiological signals and at the same time allow remote clinical surveillance through communication, processing and transmission technologies that allow to share such data everywhere.

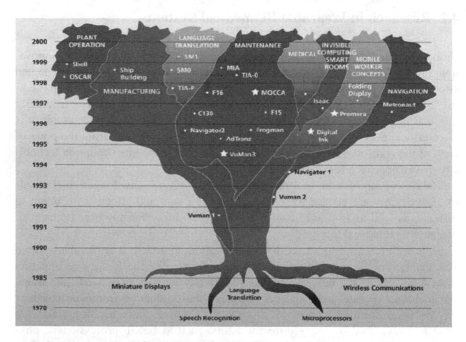

Fig. 4.1 The family tree of wearable computing technology: the medical developments and applications started in the mid-nineties in the USA

But very soon it was realized that WBS design is a complex activity [2]. First of all, on account of its multidisciplinary nature, the process of development of bio-electronic devices for diagnosis and monitoring requires knowledge and expertise from many different disciplines: Medicine, Electronics, Computer, Design, Physics, Mechanics, Chemistry and others [5]. Technological aspects such as the creation of innovative circuits, the miniaturization of electronics and its implementation on flexible substrates, are also closely linked to the issues related to wearability and usability of technology, to design, to materials engineering and chemistry but also to their characteristics and properties (washability, adhesion, sterilization,…).

Thus WBS are systems that integrate a complexity of components and technologies, each of them essential even in their extreme simplicity [3]: high technology sensors, actuators, materials, wireless communication, power control and processing unit, user interfaces, new algorithms for signal processing, chemical treatments for washability and stability of the sensors are nothing without connectors, or their proper smart positioning on the body in relation to the activity or gestures to be monitored, and the anthropometric characteristics of the subjects,

Also the proper adhesion and fitting to the body of the sensorized garments is crucial to obtain a good signal quality and minimization of artifacts during wearable monitoring: in this case a detailed analysis and study of elasticity and position of these elastic fibers in the cloth and the development of a numeric machine program for the automatic production of the garment itself are essential to obtain a good-performing wearable sensor [2].

For a better overview of the items related to WBS we can distinguish five main categories or sub-systems:

(1) hardware, i.e. the design and construction of new devices and sensors for the reliable recording of physiological signals; their physical design should also take into account shape and dimensional issues that make them ergonomic and minimally invasive (or, even better, not intrusive);
(2) system's architecture, i.e. the implementation and management of sensors' networks; for multifactorial monitoring often requires the development of systems to collect data from multiple wearable sensors and to transmit this information to the related clinical center;
(3) software, i.e. the design and implementation of innovative algorithms to extract clinically relevant information from data recorded using wearable technology;
(4) materials and industrial processes, including in this context the study and manufacture of textile materials and end products;
(5) ergonomics of each part of the system (garment, single sensor, device, software), i.e. the study and design of the physical interface of the device and its components (connectors, buttons, LEDs, displays, etc.) according to the user-centered-design approach for defining the proper physical appearance of the WBS (size, wearability, anthropometric adaptability, etc.) and cognitive study and design of the graphical interfaces of the software supporting application by the different users (patient, relatives, clinical staff).

At the Design Department of the Politecnico di Milano, a new Research team (TeDH, i.e. Technology and Design for Healthcare) started from this analysis to promote skills collaboration and integration of polytechnic disciplines [1]. In particular the SensibiLab (Laboratory of Biomedical Sensors and Systems) is the operative site and tool for the development of innovative technologies, methodologies and models for data collection, analysis and simulation of human–machine–environment physiological interaction integrating skills of medicine, engineering, design and ergonomics. The main research area of the laboratory is the development and application of Wearable Sensors and methodologies for non-intrusive measurement of biological signals associated with the spontaneous behavior of the subject [6] (Fig. 4.2).

This technological and methodological know-how of the Sensibilab has not been kept for research only, but gave rise to specific IPR from which two new business initiatives, each one with a different identity and specificity, were established.

4.3 From Research to Market: IPR as a Key Element to Code and Protect the Developed Innovation

Among its institutional goals, the Politecnico di Milano devotes special attention and commitment to the promotion and development of research and technology transfer.

The two main tools that the University utilizes for this purpose are:

(1) the protection of the results of research carried out within the University;
(2) the transfer of technology related to the intellectual creations of its researchers.

Fig. 4.2 The prototype of wearable sensor and of monitoring unit for newborns developed at the SensibiLab of the Politecnico di Milano

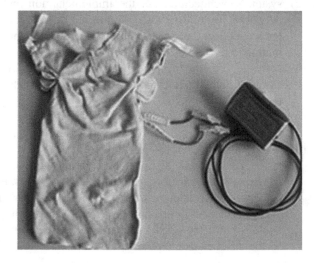

The direct transfer of technology from universities to industry can be accomplished in different forms: (a) patent licensing agreements, and/or (b) joint patents with a company, or (c) creation of joint applied research laboratories and testing procedures.

They are complementary and they probably represent the more standard option and, traditionally, the most used form.

But by developing among academics, researchers, and graduate students an innovative culture of enterprise and entrepreneurship, it is possible and desirable that they can take an active part in the creation of new businesses. In this way, the creation of university spin-off and start-up is not only the result of innovative technology originated in the university, but also the ability to create a group of individuals working directly to evolve technology/prototypes into new products, systems or services, with the possible support of one or more industrial partners, [7].

The idea behind this proposal is that ownership of the innovation can ease translation into new products.

In this context, Politecnico di Milano is promoting initiatives by its facilities, staff and students for the creation of new companies (for the production of goods or services) arising from the research results of the University and therefore consistent with the academic purpose of exploitation of research [8].

In general, a university *spin-off* is a new company that brings to the market the results and technologies developed through research by offering services and products, particularly in high technology sectors.

Such an initiative is in a favorable condition for entrepreneurial innovation by virtue of the close link with the research world and because, on account of its approach/culture, it invests a considerable part of its resources in R&D.

Typically, a spin-off company includes among its founding members the University that has generated the innovation on which the company know-how is based. This means that the University is a shareholder.

There are however cases where the University is not directly involved but the company implements an innovative product derived from university research, for instance under a patent license agreement.

This definition previously identified the academic *start-ups*, i.e. any business company in which the University does not participate directly as a shareholder but which aims at the development of pre-competitive products and services derived from the research results of the University itself.

The creation of a spin-off or of a start-up that can have real potential for success on the market is based on some essential requirements:

- the *technology*, defined as the set of knowledge produced by the University and responding in a unique and meaningful way to a market need elicited from a clear and rigorous analysis that identifies strengths and weaknesses, as well as risks and opportunities in the short and especially in the long term; this knowledge can be either codified (e.g. in a formula, described and commented for understanding) or simply practical, protected by IPR (e.g. in a patent). This point can be summarized in two parts: the patent or the uncodified know-how and a draft business plan (that is to be completed only through an analysis carried out by experts in the field and preferably not by the holders of the know-how, who are often positively biased).

- an *industrial partner*, whether or not participating in the shareholder structure, with the function to exit the limbo of the laboratory prototype and start the porting to the industrialization of the know-how of research;
- a *business partner*, in this case preferably to be committed and therefore being part of the company shareholders; it has the knowledge and ability to penetrate the market, with expertise in that specific area, it is well-structured and it has a very different experience from that of academic researchers;
- *human resources* that are integrated and complementary in terms of human and professional skills; thanks to the balance of their own (technical, production, commercial, organizational and management) abilities and specificity, they can develop a consistent business plan; everyone must be aware of the importance to take care of and be responsible for what they do and nobody must expect to invade or steal the roles of others (technicians must do technical jobs, sales staff must do the selling, managers must organize and manage,...);
- *financial resources*, which are used for the completion of the development phase, the phase of industrialization of the product/service, and its introduction on the market.

This list should not and does not intend to be a default 5-rule mandatory list, but, with the individual and appropriate customization to be assessed case by case, it is believed that this "basic recipe" can determine the success of the business project.

Another important variable to be considered in the creation and management of a spin-off is *time*. Many industrial aspects are related to this parameter, so simple yet so critical:

- *The time-to-market*. The know-how or patent cannot and must not remain too long a prototype but it has to be industrialized, go out and face the market, sometimes the first who arrives in the market takes a significant competitive advantage, sometimes it also induces the need in the market and thereby determines the success of the initiative;
- *The hurry*. Notwithstanding the foregoing, it is also important not to be too much ahead of one's time: if a product is not mature and ready but is marketed anyway, a very possible risk is that it can create frustration and negative feedback; thus, even if the product is very promising, it can destroy the opportunity. Also the hurry to disclose a new product too early with respect to its availability on the market is a factor that may be critically negative.
- *The patience*. A common and popular saying states that to build a solid and mature company (and product) it takes at least three years. This means that the target time perspectives (and the expectation to have a positive economic return) cannot be in the immediate period, but must be reasonably placed in a time scale from medium to long term.

These considerations represent the synthesis of a set of lessons-learned during the spin-off actions at the Politecnico di Milano.

Now two case-studies are described: one spin-off and one start-up both implementing wearable biomedical systems.

4.4 Wearable Monitoring Systems. Case Study 1: SXT—Sistemi per Telemedicina s.r.l

In the first phase, the research carried out in the SensibiLab focused on the development of electronic technologies for wearable biomedical systems: the goal was to find innovative unconventional solutions for the acquisition and processing and transmission of vital signs that can be obtained through wearable sensors.

The outcome has been an innovative analog front-end for the detection of bioelectric signals, characterized by remarkable performance in terms of speed of response, quality and stability of the signal. This innovation has been patented by the Politecnico di Milano: the patent has been internationally granted (Fig. 4.3).

Then, the University had to decide the strategy of exploitation together with the inventors. Two opportunities were selected: (a) licensing, (b) direct exploitation by founding a spin-off company. This second option was chosen for two main reasons: (1) the opportunity for the inventors (researchers and PhD fellows) to be directly involved in the industrial exploitation of their work, (2) the possibility to create new jobs for young researchers of the University to keep them close to university (even if outside of it) together with the important skills they have developed and that can be of possible interest for future research initiatives. This proximity also allows the rapid transfer of new know-how that ongoing research is continuously developing. With this solution, research can also take advantage of an industry that supports the research itself and at the same time the industrial implementation of the new services and products.

SXT—Sistemi per telemedicina Srl was born as a spin off of the Politecnico di Milano in June 2006.

In the same period it won a Business Plan competition of the Lecco Province for the support of technological innovation of SMEs and in the development of new business initiatives in the Province Area: the proposed Business Plan envisaged the creation of a set of monitoring devices for the exploitation of telemedicine which is strategic for the Lecco Province, whose territory is geographically complex. Thus SXT was proposing to be the technological partner for telemedicine developments thanks to the industrial production of the latest innovations in the context of wearable biomedical technologies. This funding has provided the financial resources needed for the first-year activity.

In the company, one senior assistant professor and three young researchers who obtained their Ph.D in Biomedical Engineering at the Politecnico di Milano decided to participate in the SXT adventure. They worked in the Laboratory where the development of the know-how in WBS took place, and now they are leading the porting of such technological expertise into the products and systems of SXT.

(19)   [logo: Europäisches Patentamt / European Patent Office / Office européen des brevets]

[barcode]

(11)   **EP 1 977 505 B1**

(12)   **EUROPEAN PATENT SPECIFICATION**

(45) Date of publication and mention   (51) Int Cl.:
     of the grant of the patent: *H03F 1/34* (2006.01)   *H03F 3/183* (2006.01)
     06.03.2013 Bulletin 2013/10   *H03F 3/45* (2006.01)

(21) Application number: 07702887.6   (86) International application number:
                                                    PCT/EP2007/000456
(22) Date of filing: 19.01.2007
                                               (87) International publication number:
                                                    WO 2007/085383 (02.08.2007 Gazette 2007/31)

(54) **SIGNAL CONDITIONING CIRCUIT**

      SIGNALKONDITIONIERUNGSSCHALTUNG

      CONDITIONNEUR DE SIGNAL

(84) Designated Contracting States: • PICCINI, Luca
     AT BE BG CH CY CZ DE DK EE ES FI FR GB GR   I-20137 Milano (IT)
     HU IE IS IT LI LT LU LV MC NL PL PT RO SE SI   • MAGGI, Luca
     SK TR   I-24121 Bergamo (IT)

(30) Priority: 24.01.2006 IT MI20060113   (74) Representative: Ciceri, Fabio et al
                                                    Perani & Partners
(43) Date of publication of application: Piazza San Babila, 5
     08.10.2008 Bulletin 2008/41   20122 Milano (IT)

(73) Proprietor: Politecnico Di Milano   (56) References cited:
     20133 Milano (IT)   US-A- 2 033 330   US-A- 3 972 002
                                                     US-A- 4 189 681   US-A- 4 257 006
(72) Inventors: US-A- 5 444 418
     • ANDREONI, Giuseppe
       I-20053 Muggio' (IT)

Printed by Jouve, 75001 PARIS (FR)

EP 1 977 505 B1

Fig. 4.3 The patent describing an original architecture for biosignal recording specifically dedicated to WBS by the researchers of the Politecnico di Milano

(12) INTERNATIONAL APPLICATION PUBLISHED UNDER THE PATENT COOPERATION TREATY (PCT)

(19) World Intellectual Property Organization
International Bureau

(43) International Publication Date
2 August 2007 (02.08.2007)

PCT

(10) International Publication Number
WO 2007/085383 A1

(51) International Patent Classification:
H03F 1/34 (2006.01)

(21) International Application Number:
PCT/EP2007/000456

(22) International Filing Date: 19 January 2007 (19.01.2007)

(25) Filing Language: English

(26) Publication Language: English

(30) Priority Data:
MI2006A000113 24 January 2006 (24.01.2006) IT

(71) Applicant *(for all designated States except US)*: **PO-LITECNICO DI MILANO** [IT/IT]; Piazza Leonardo da Vinci, 32, I-20133 Milano (IT).

(72) Inventors; and
(75) Inventors/Applicants *(for US only)*: **ANDREONI, Giuseppe** [IT/IT]; Via Italia, 93, I-20053 Muggio' (IT). **PICCINI, Luca** [IT/IT]; Via Ciceri Visconti Laura 14, I-20137 Milano (IT). **MAGGI, Luca** [IT/IT]; Via Vittorio Emanuele, 40/B, I-24121 Bergamo (IT).

(74) Agents: SIMINO, Massimo et al.; c/o Perani Mezzanotte & Partners, Piazza San Babila, 5, I-20122 Milano (IT).

(81) Designated States *(unless otherwise indicated, for every kind of national protection available)*: AE, AG, AL, AM, AT, AU, AZ, BA, BB, BG, BR, BW, BY, BZ, CA, CH, CN, CO, CR, CU, CZ, DE, DK, DM, DZ, EC, EE, EG, ES, FI, GB, GD, GE, GH, GM, GT, HN, HR, HU, ID, IL, IN, IS, JP, KE, KG, KM, KN, KP, KR, KZ, LA, LC, LK, LR, LS, LT, LU, LV, LY, MA, MD, MG, MK, MN, MW, MX, MY, MZ, NA, NG, NI, NO, NZ, OM, PG, PH, PL, PT, RO, RS, RU, SC, SD, SE, SG, SK, SL, SM, SV, SY, TJ, TM, TN, TR, TT, TZ, UA, UG, US, UZ, VC, VN, ZA, ZM, ZW.

(84) Designated States *(unless otherwise indicated, for every kind of regional protection available)*: ARIPO (BW, GH, GM, KE, LS, MW, MZ, NA, SD, SL, SZ, TZ, UG, ZM, ZW), Eurasian (AM, AZ, BY, KG, KZ, MD, RU, TJ, TM), European (AT, BE, BG, CH, CY, CZ, DE, DK, EE, ES, FI, FR, GB, GR, HU, IE, IS, IT, LT, LU, LV, MC, NL, PL, PT, RO, SE, SI, SK, TR), OAPI (BF, BJ, CF, CG, CI, CM, GA, GN, GQ, GW, ML, MR, NE, SN, TD, TG).

Declaration under Rule 4.17:
— *of inventorship (Rule 4.17(iv))*

Published:
— *with international search report*
— *before the expiration of the time limit for amending the claims and to be republished in the event of receipt of amendments*

[Continued on next page]

(54) Title: SIGNAL CONDITIONING CIRCUIT

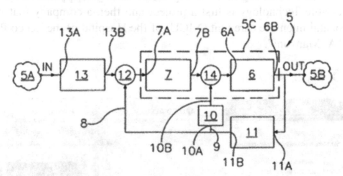

(57) Abstract: The present invention relates to a signal conditioning circuit, comprising: an amplification and filtering chain (5C) having amplifier means (6) and filter means (7), each with an input connection (6A, 7A) and an output connection (6B, 7B); a first feedback path (8) for providing feedback to said amplification and filtering network (6, 7), including a first phase-shifting network (11) having an input connection (HA) and an output connection (11B). The circuit (5) is characterized in that it comprises a second feedback path (9) for providing feedback to said amplifier means (6), including a second phase-shifting network (10) having an input connection (10A) and an output connection (10B); said input connection (11A) of said first phase-shifting network (11) being connected with said output connection (6B) of said amplifier means (6); said output connection (HB) of said first phase-shifting network (11) being connected with the input connection (10A) of said second phase-shifting network (10) and with said input connection (7A) of the filter means (7); said first phase-shifting network (11) being designed to compensate for errors in said output connection (6B) of said amplifier means (6) or to operate when said output connection (6B) of said amplifier means (6) is in a saturated condition.

Fig. 4.3 (continued)

Together with them, the shareholding structure required since the beginning an industrial partner with the purpose of supporting research in the process of know-how industrialization and transformation into a product to be launched on the market, and of performing the role of the leading actor in the business.

SXT is characterized by the R&D attitude to develop technological and methodological innovation to be up-to-date in the market sector where it is positioned in and that is characterized by excellent growth prospects.

Today (August 2013) SXT is ISO13485 certified and offers, either directly or in co-branding, a variety of products, specifically compact and user-friendly systems for remote monitoring of a plurality of biological signals (polygraphs). It also offers technical support, solutions design and industrialization of new products and systems for SMEs and multinational companies that require such skills.

The latest product STX developed is Cardiodial, an innovative and miniaturized single-lead ECG with GPRS data transmission for assistance in mobility, anywhere (Fig. 4.4).

4.5 Wearable Monitoring Systems. Case Study 2: ComfTech (Comfortable Technology) s.r.l

The history of ComfTech is the outcome of a meeting, a meeting of people, a meeting of skills, a meeting of technologies, a meeting of opportunities.

Comfortable Technology is first a project and then a company that was set up in the neonatal intensive care unit (NICU) of the Hospital of the Lecco Province—Presidio "A. Manzoni".

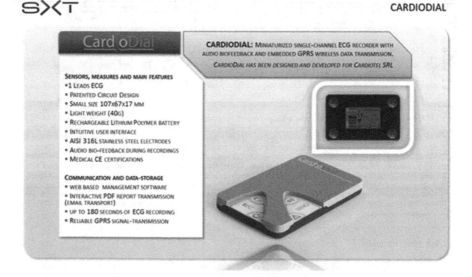

Fig. 4.4 The cardiodial system by SXT

By chance, on the same day and at the same time, one researcher from Sensibilab and one fashion designer specialized in baby clothes design independently went to the director of the NICU: Sensibilab offered the possibility to monitor vital signs of premature infants using textile sensors, and the textile company proposed the development of a line of clothing for premature babies. From this complementary know-how about:

- biomedical technologies, in particular for the non-intrusive measurement of physiological parameters;
- textiles, specializing in the design for babies, from premature newborns to adolescents, and technologies for the production of textiles and textile articles in general;
- ergonomics and Design for usability, wearability, comfort and acceptance of the system by the user,

the first steps of a synergy of purpose and energies were taken.

The original idea/product of ComfTech was studied closely with doctors and nurses, to design all of its components to converge into a product that wanted to be highly technological but at the same time of absolute simplicity and ease of use for both the users, first and foremost newborns, and secondly family members and medical and paramedical personnel who have to deal with the child daily.

The first goal was the development of a reliable but extremely flexible (in terms of applications) textile sensors' system that could be directly connected to the existing standard monitors for NICU, or that could independently manage data from home through a dedicated monitoring unit. This implied the creation of an absolutely comfortable textile sensor, that should be also easy to maintain, and especially of a garment that, due to its shape and modeling, could perfectly implement the "Care"[1] in pre-term newborns. This direct experimentation and the development of a product designed in close contact with doctors and nurses was also a key market strategy to encourage the early adoption of the system into clinical practice.

The main advantages of such a sensorized garment for newborns (also premature ones) are:

- the wearable monitoring does not prevent or restrict the natural contact of the mother with the child (bonding), that is considered more crucial than the required monitoring;
- thanks to its shape and modeling, the garment provides the feeling of "containment" of the newborn, simulating the uterine sack, which is considered an important practice for the "Care" approach in premature newborns;

---

[1] Specifically, the term "Care" designates the approach of doctors and nurses in the care of newborns in the neonatal intensive care unit (and in any case in the first hours of life) aimed at ensuring the child the most pleasant and comfortable physical experiences, in order to offer the best conditions for proper development.

- the wearable sensor system does not prevent or limit the standard operation of doctors and nurses;
- the textile system as a component/sensor is able to "live" independently and communicate with standard hospital monitors (different monitors of different companies were tested) and can therefore be integrated in ordinary hospital systems, without involving procedural changes.

Instead, the home version has to be integrated with a hardware unit and a corresponding software suite. In this option the complete system can manage the data independently, report them in "readable" form for non-professionals, and eventually store them for a possible reinterpretation by the medical and paramedical staff.

This idea and prototype led to a patent application to be first submitted to the judgment of a selected panel of experts, entrepreneurs and venture capitalists. Under suggestion of the University, ComfTech took part in a pre-Business Plan competition, namely the I2P—"Idea to Product Competition Italy", organized for the first time in 2010 by the Business Accelerator of the Politecnico di Milano and the Fondazione Politecnico di Milano.

This competition aims to bring out research ideas that can be turned into products. The final of the 2010 edition was held on 8 June 2010. ComfTech ranked fourth and thus was admitted to a tutoring service for the preparation of a complete Business Plan.

This opportunity allowed to perform a detailed analysis of the position and opportunities of such a concept in the market concerning solutions for children monitoring, the technological feasibility and all the industrial aspects, in particular the organizational and financial ones.

The Business Plan was then presented at the 2011 edition of the Start Cup Milano Lombardia, whose final was held in Milan on 26 October 2010.

Here the ComfTech project won:

- the 1st Prize in the Life Sciences Section;
- the Special Prize for the best business idea of the Lecco Province;
- the special mention of the "Bright Future Ideas Award" of the British Consulate General and UK Trade & Investment Milan.

By virtue of this result, Comfortable Technology was selected and admitted to the National Innovation Award PNICube where it was selected for the final as one of the 10 best business ideas in Italy in 2010.

These successes strengthened the conviction of the entrepreneurship idea and provided the initial financial input to the company.

The prototype was patented in co-ownership between the Politecnico di Milano (Fig. 4.5) and the new company, that obtained an exclusive license.

Comfortable Technology (ComfTech) was set up to create products that improve the quality of life of fragile users in terms of comfort and usability of technology, associating the technological aspect of miniaturized systems to the textile world, that has always offered protection and physical and psychological

(12) INTERNATIONAL APPLICATION PUBLISHED UNDER THE PATENT COOPERATION TREATY (PCT)

(19) World Intellectual Property
Organization
International Bureau

(43) International Publication Date
7 June 2012 (07.06.2012)

WIPO|PCT

(10) International Publication Number
WO 2012/073076 A1

(51) International Patent Classification:
A41D 13/12 (2006.01)   A61B 5/026 (2006.01)
A61B 5/024 (2006.01)

(21) International Application Number:
PCT/IB2011/000671

(22) International Filing Date:
29 March 2011 (29.03.2011)

(25) Filing Language: Italian

(26) Publication Language: English

(30) Priority Data:
MI2010A002245 3 December 2010 (03.12.2010)   IT

(71) Applicants (for all designated States except US):
COMFTECH S.R.L. [IT/IT]; Via Castello, 9, I-20900
Monza (MB) (IT). POLITECNICO DI MILANO
[IT/IT]; Piazza Leonardo da Vinci, 32, I-20133 Milano
(MI) (IT).

(72) Inventors; and
(75) Inventors/Applicants (for US only): ANDREONI, Giuseppe [IT/IT]; Via Italia, 93, I-20835 Muggiò (MB) (IT).
MOLTANI, Lara, Alessia, Laura [IT/IT]; Via Castello, 9,

I-20900 Monza (MB) (IT). ZANINI, Rinaldo [IT/IT]; Via
Tubi, 12, I-23900 Lecco (LC) (IT). BELLU', Roberto,
Carlo [IT/IT]; Via Bergamo, 26, I-20038 Seregno (MI)
(IT).

(74) Agents: CICERI, Fabio et al.; c/o Perani & Partners,
Piazza San Babila, 5, I-20122 Milan (IT).

(81) Designated States (unless otherwise indicated, for every
kind of national protection available): AE, AG, AL, AM,
AO, AT, AU, AZ, BA, BB, BG, BH, BR, BW, BY, BZ,
CA, CH, CL, CN, CO, CR, CU, CZ, DE, DK, DM, DO,
DZ, EC, EE, EG, ES, FI, GB, GD, GE, GH, GM, GT, HN,
HR, HU, ID, IL, IN, IS, JP, KE, KG, KM, KN, KP, KR,
KZ, LA, LC, LK, LR, LS, LT, LU, LY, MA, MD, ME,
MG, MK, MN, MW, MX, MY, MZ, NA, NG, NI, NO, NZ,
OM, PE, PG, PH, PL, PT, RO, RS, RU, SC, SD, SE, SG,
SK, SL, SM, ST, SV, SY, TH, TJ, TM, TN, TR, TT, TZ,
UA, UG, US, UZ, VC, VN, ZA, ZM, ZW.

(84) Designated States (unless otherwise indicated, for every
kind of regional protection available): ARIPO (BW, GH,
GM, KE, LR, LS, MW, MZ, NA, SD, SL, SZ, TZ, UG,
ZM, ZW), Eurasian (AM, AZ, BY, KG, KZ, MD, RU, TJ,
TM), European (AL, AT, BE, BG, CH, CY, CZ, DE, DK,

[Continued on next page]

(54) Title: WHS ITEM OF CLOTHING FOR DETECTION OF VITAL PARAMETERS OF A BABY

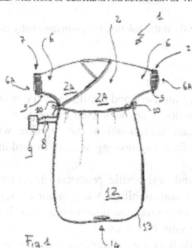

(57) Abstract: The present invention relates to a WHS item
of clothing for a baby, said WHS item of clothing being
configured for detection of vital parameters of the baby, said
WHS item of clothing comprising a first portion adapted to
at least partially cover the torso of the baby and at least one
sensor for detecting a vital parameter of the baby. It is characterized in that it comprises at least one second portion ex-
tending from the first portion, said second portion being adapted to at least partially cover one arm of the baby, said at
least one second portion having a section that can be
stretched in at least one direction, said stretchable section
being designed to remain in a fixed position of said at least
one arm; said at least one sensor being coupled in said
stretchable section to be properly positioned relative to the
baby's arm, to provide an electrical output signal indicative
of the vital parameter being detected; a connecting cable op-
erably connected with said at least one sensor to transmit
said electrical signal to a data collection unit.

Fig. 4.5 The joint patent application by ComfTech and Politecnico di Milano concerning a
wearable sensor system for monitoring premature babies

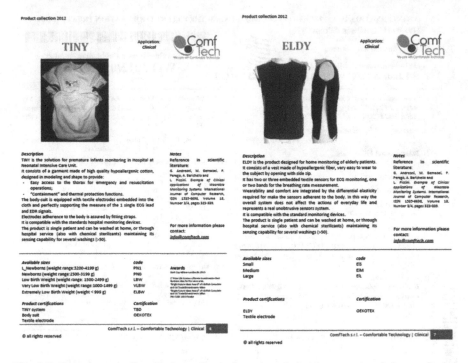

Fig. 4.6 Some items in ComfTech catalogue: from the solution for premature baby monitoring to sensorized garment for the elderly

comfort. The combination of technology industry and textile industry is still little explored, mainly in the sense that technology has been used to create new textile products (yarn, fabric). Instead in this case the rationale is the opposite, where the textile technology does support biomedical engineering factors in making more comfortable and user-friendly systems.

Therefore the goal is to produce and sell textile products characterized by highly innovative textiles, functionalities and usability. A key factor is ergonomics and design, in terms of smart sensors integration and positioning, usability, wearability, comfort and acceptance of the system by the user.

In this sense we can think of ComfTech as a complementary initiative to SXT. Thanks to their independence, they can also find different opportunities that can lead to mutual synergies.

Today ComfTech has a catalog covering sensorized systems and garments for a wide range of applications and for all age groups (Fig. 4.6). The company began the procedure for certification, that is expected to be obtained by the end of summer 2013.

In the same period, the first system for home monitoring of newborns will be released on the market.

4.6 Conclusions

e-Health is characterized by a strong, strategic and unique opportunity: the demand for health is steady, or better it is even increasing anytime, anywhere. The products that address growing transnational needs (aging of the population, security and sports) can be an extraordinary business opportunity.

The ease of use of wearable biomedical systems and sensorized clothing makes them a very attractive solution with many advantages.

SXT and ComfTech are proposing themselves in this area as strategic partners, each one with specific but complementary skills and products. This allows both of them to operate with the maximum freedom of action in the market but without overlapping.

The presence of spin-off in the production structure of a country is an invaluable resource to maintain a high level of competitiveness in the global market, where countries with high production potential and low labor costs make unsustainable competition in industrial LowTech frames. Technological innovation, including the spin-offs that are its result and its engine at the same time, is the indispensable condition to support industrial development based on the excellence of the products in competitive markets. Furthermore, this process produces a positive feedback on the academic system with support for research with new funds and employment opportunities for new graduates and PhDs.

By virtue of their small size, another important element that these units can pursue is speed of response in proposing new systems, solutions or products. This may allow to hold and, where possible, to manage from a strong position some high technology areas of the market. This latter aspect is for both companies a strategic and crucial goal.

References

1. Andreoni G (2011) Progettare la salute integrata, DA, pp 47–49
2. Andreoni G (2008) Sistemi di sensori indossabili per il monitoraggio: dalla Ricerca al Mercato. In: Bonfiglio A, Cerutti S, De Rossi D, Magenes G (eds) Sistemi indossabili intelligenti per la Salute e la Protezione dell'Uomo. Patron, pp 357–384
3. Andreoni G, Bernabei M, Perego P, Barichello A, Piccini L (2011) Example of clinical applications of wearable monitoring system. Int J Comput Res 18(3/4):323–339
4. Andreoni G, Barbieri M, Piccini L (2011) A review of the intellectual property right in the field of wearable sensors and system. Int J Comput Res 18(3/4):269–285
5. De Rossi D, Lymberis A (2005) Guest Editorial, Special section on new generation of smart wearable health systems and applications. IEEE Trans Inf Tech Biomed 9(3):293–294
6. Romero M, Perego P, Andreoni G, Costa F (2010) New strategies for technology products development in HealthCare. In: Meng Joo Er (ed) New trends in technologies: control, management, computational intelligence and network systems. Sciyo, pp 131–142
7. Bassi D (2011) Trasferimento tecnologico dall'Università all'impresa. In: Knowtransfer 1(1) http://knowtransfer.unitn.it/1/trasferimento-tecnologico-dall-universita-all-impresa. Accessed 17 Jan 2013

8. Politecnico di Milano (2011) Regolamento Spin–Off di Ateneo, pp 1–8
9. Patel S, Park H, Bonato P, Leighton C, Rodgers M (2012) A review of wearable sensors and systems with application in rehabilitation. J Neuro Eng Rehabilit 9:21 http://www.jneuroengrehab.com/content/9/1/21

Web References

http://www.sxt-telemed.it
http://www.comftech.com
http://www.epo.org

Index

G. Andreoni et al., *Developing Biomedical Devices*, PoliMI SpringerBriefs,
DOI: 10.1007/978-3-319-01207-0, © The Author(s) 2014